THE
WORLD
FOOD
CRISIS

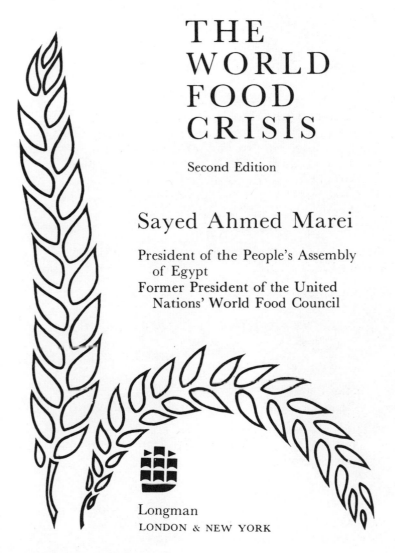

THE
WORLD
FOOD
CRISIS

Second Edition

Sayed Ahmed Marei

President of the People's Assembly
of Egypt
Former President of the United
Nations' World Food Council

Longman
LONDON & NEW YORK

LONGMAN GROUP LTD
LONDON

*Associated companies, branches and
representatives throughout the world*

First published 1976
Second Edition 1978

ISBN 0 582 78077 2

Library of Congress Cataloging in Publication Data

Marei, Sayed.
   The world food crisis.

   Includes index.
   1. Food supply.   2. Food supply—Arab countries
I. Title.
HD9000.5.M338   1977   338.1′9   77-14527
ISBN 0-582-78077-2

Printed in Great Britain by
Butler & Tanner Ltd, Frome and London

Contents

"The very first day that men opened their eyes in the world they knew hunger and the use of food."

Corestia et Fame by
Giovanni Battista Segni
(Bologna 1602), p. 115

Preface

It is a truism to say that the world is getting smaller every day. It is equally true to say that man's achievements and man's institutions are growing greater every day. So are his problems.

Fifty years ago, the country with a population of thirty million was considered big; the country with a million cultivated acres was considered productive; the country with a hundred or so factories was considered advanced; the country having a university or two was considered cultured; and the country whose wealth totalled a thousand million dollars was considered rich.

Now, all this is changed. Today, a nation possessing these features would rank perhaps the hundredth among the nations of the world.

The present epoch is moving more and more towards bigness. Indeed, it is an epoch characterized by big groupings, interacting with one another. Historically, we see a world where first there was one family against another; then a tribe against another tribe, a city against another city; then a nation against another nation. But now we see groups of nations opposed to other groups.

In the year 1850, there were four cities in the whole world whose population exceeded a million. A century later, the figure had risen to 145. If such growth continues, the number of such cities will double in no more than ten years.

What is important is that this change is not a change in size alone, but in the manner and standard of life and in the degree of power.

For these reasons, it no longer seems appropriate to classify people solely according to race or country or creed; classification must be in terms of where they stand in relation to the mode of life in this epoch: life in the twentieth century, and perhaps also life in the twenty-first century.

Belonging to the twentieth century implies development, education, science, technology, industrialization, and unity. Five hundred years ago, unity was found only among peoples of one religion or one creed. Later, the basis of unity was a common foe; and then a common ideology. But now, the unity that we see is established for some common advantage.

It is the mutual benefit from progress and from keeping pace with advance that has resulted in unity between France and Germany, after a modern history replete with bloodshed; it is the conception of mutual benefit that now induces the United States of America to erect factories in the Soviet Union after years of division arising from ideological warfare. These changes should not be construed to mean that the foes of yesterday are suddenly the friends of today. No, it means that there is a bond, a simple and easy bond at that, that links one to the other. It is as if one says to the other: "We are two individuals living in this world. You have your home and I have mine. You have your affairs and I have mine. You have a hundred objectives and so have I. Let us take just one of these objectives and work together towards its realization. This objective may be the production of a certain make of aeroplane or an innovation in printing processes. Either one of us is capable of achieving this in ten years' time, but if we co-operate the years will be reduced to one."

This is a simplified illustration of the manner in which the advanced countries begin their common economic ventures. It is a method aimed at achieving results, not at creating differences. Differences will continue, but the achievements will reduce many differences.

It is easy, then, to see how the world today may overcome its economic problems and can continue its progress. But in spite of all the advances made, we suddenly find ourselves facing a problem which is as old as mankind but which, like other facets of society, has grown to an enormous size: the problem of hunger. In the last few years, countless thousands have died from undernourishment. Now, famine threatens millions. A life of squalid poverty is the lot of even more. The danger of starvation grows as progress grows: at an exponential rate.

Are we to say that a world which can send a man to the moon, which can cure the majority of diseases, and can produce machines to solve the most complex mathematical problems within seconds can still deny to millions of its inhabitants sufficient basic food to keep them alive? It sounds incredible. In fact, it seems probable.

That is the topic of this book. I believe that solutions are possible, and I suggest what they may be. Because I am an Arab, I have used the Arab World to illustrate in detail how such solutions can be achieved. We in the Arab World have economic problems which are themselves examples of the plight of the rest of the world: we have our rich nations and our poor nations, our rich individuals and our poor individuals; we have land and resources which we have not yet learned to exploit to the full; we have money that is poorly used and poverty which we have not overcome; we have failed to establish the co-operation and the unity which could bring real success.

And this is the situation of the rest of the world, too. No one needs to be convinced that it is an intolerable situation; we all know that. What we need to do now is to resolve it.

Preface to
Second Edition

Since the appearance in 1976 of the first edition of this book, some changes have taken place in the immediate food situation. At world level, production increased by an average of 3 percent, the biggest increase since 1973. Cereal production rose by 9 percent to a new record level. But averages can be misleading. Food production in the MSA (Most Seriously Affected) countries has increased at a rate of only about 2 percent per year in the seventies, showing an actual decline in their production *per capita*. Even the world figure of 3 percent is still short of the annual growth target of 4 percent agreed upon by the World Food Conference.

I must therefore warn that the challenge facing us today is greater than it was in 1974. Any complacency would show a misinterpretation of the facts, and would eventually lead to a crisis of larger proportions. It is imperative that we recognise the full dimensions of the food problem, with its social, economic, and political implications, in the realisation that a solution is a most important element of a New International Economic Order.

The demands on the world community for co-operation are now more, not less, urgent than they were two or three years ago.

May 1977

Part
One

The Crisis

Chapter One

The Crisis

Since the Second World War, a new branch of economics has developed: Growth Economics. The post-war period has seen a proliferation of growth theories and analyses of the determinants of growth. Questions were asked why many industrial countries have been experiencing full employment[1] since the war. This constituted a marked departure from neo-classical economics and even from Keynesian economics, the latter usually considered to be modern economics. Economists carried out many exercises to dynamize Keynes's short-run theories and turn them into long-run theories of growth in the advanced countries.

In the meantime, there developed another new branch in economics which laid emphasis on the poor countries: Development Economics. Since the early fifties, scores of books were published under different titles. These books attempted to delineate certain characteristics of the then so-called underdeveloped countries.[2] Some writers tried to apply the growth theories of the time to these countries;

---

1 Full employment is used here in the usual sense, which often includes a small percentage of unemployment, mainly frictional unemployment (i.e. unemployment as a result of changing jobs).
2 Since the fifties, those countries have been referred to as underdeveloped, or poor, or developing or development economies, the latter expression implying that development is the problem for these countries. Recently the term "less developed" has been used.

others concentrated on outlining solutions to some key obstacles, such as export instability, capital deficiency, population pressures, and so forth. After a time, more and more emphasis was laid on the socio-economic institutions.

Be that as it may, the emphasis was on growth. Until the beginning of the present decade the economies of the advanced countries enjoyed a fair degree of monetary stability, price stability, and growth. The long depression of the late nineteenth century appeared in the books of economic history as a problem that could now be easily surmounted with the modern tools of economics. And the same was said of the hyper-inflation that occurred in Germany after the First World War, and the Great Depression of the thirties. It was shown that modern economics could easily achieve full employment without fluctuations, without inflation or depressions.

The post-Second World War period was also characterized by major strides in technology and science. Inventions and innovations were always available to lift an economy from the dismal stationary state of zero growth which the classical economists had often forcefully depicted.

Paradoxical as it may appear, in 1973 the world woke up rather suddenly to a problem or a crisis, the implications of which seemed to transcend national boundaries – and it is no wonder that it was referred to as the *world* food crisis. In a world of science, technology, and inventions, the food issue appeared such a simple and commonplace issue, the solution of which was certainly much simpler than other matters. In other words, such pessimists as Malthus[1] who warned about the danger of population growth outstripping food supply seemed to have no place in our present-day world. Difficult as it may be to believe, the food crisis is still

---

1 It is customary to refer to Malthus as one who painted a gloomy world. This is somewhat unfair to the writer. If one reads the first four pages of the first edition of the *Essay on Population*, the picture is clearly gloomy, where population increases in a geometrical progression and food supply in an arithmetic progression. But in subsequent editions, Malthus modi-

with us. Before I go into some of the reasons for the crisis, I will attempt to outline here its dimensions.

The estimates which were prepared for the World Food Conference show that in developing countries some 460 million people, of whom almost half are children, are undernourished. Words such as malnourishment, or undernourishment, or even famine, have developed in common usage vague and secondary meanings. It is not easy to present a precise picture. Table 1 gives a rough indication of the dietary energy supply and protein supply in each of the countries concerned.

In preparing these figures, use was made of the Report of a Joint FAO/WHO *Ad Hoc* Expert Committee, FAO, Rome, 1973. But it should be understood that despite the progress in definition and evaluation of undernourishment, there are some factors which are masked by the statistics. For example, if a diet is adequate in its protein content but the actual quantity of food eaten provides less energy than is needed, some of the protein will be used mainly as a source of energy, and consequently not be fully utilized for its protein functions. In addition, the data are averages at the national level. Unequal distribution of available supplies in any one country aggravates the problem.

Even with all these reservations in mind, the picture is quite dismal. The developing countries comprise two-thirds of the world's population and are increasingly relying on imports to satisfy their needs. This increased dependence on food imports (largely cereals) is a new phenomenon. In the late thirties, Latin America, Eastern Europe, the USSR, Africa, Asia, Australia and New Zealand were all grain exporters. Western Europe was the big importer

---

fied his views somewhat when he allowed for what he calls "melioration of the land". In any case, Malthus was drawing attention to a real resource problem versus population, the latter having a very strong power to go on increasing.

Population, food supply and demand for food in individual countries

Table 1

	Population	Food production[1]	Demand for food[2,3]	Energy supply[3,4]		Protein supply[3,4]
	Percentage rate of growth per year[5]			Calories per caput per day	Percentage of requirement[6]	Grammes per caput per day
Developed Countries						
Albania	2·8	3·6	4·6	2,390	99	74
Australia	2·1	3·7	2·4	3,280	123	108
Austria	0·4	2·5	1·1	3,310	126	90
Belgium–Luxembourg	0·6	2·1	1·2	3,380	128	95
Bulgaria	0·8	4·3	2·8	3,290	132	100
Canada	2·2	2·2	2·5	3,180	129	101
Czechoslovakia	0·9	1·8	1·9	3,180	129	94
Denmark	0·7	1·6	1·3	3,240	120	93
Finland	0·8	2·4	1·1	3,050	113	93
France	1·0	3·0	2·0	3,210	127	105
German Dem. Rep.	0·3	1·6	0·8	3,290	126	87
Germany, Fed. Rep.	1·0	2·5	1·9	3,220	121	89
Greece	0·8	4·0	2·3	3,190	128	113
Hungary	0·5	3·0	1·9	3,280	125	100
Ireland	0·1	1·7	0·3	3,410	136	103
Israel	3·4	7·7	4·9	2,960	115	93
Italy	0·7	2·9	2·3	3,180	126	100

Japan	1·1	4·3	3·7	2,510	107	79
Malta	0·1	3·2	1·2	2,820	114	89
Netherlands	1·3	3·0	1·7	3,320	123	87
NewZealand	2·1	2·7	2·0	3,200	121	109
Norway	0·9	1·3	1·3	2,960	110	90
Poland	1·4	3·0	2·3	3,280	125	101
Portugal	0·6	1·7	2·3	2,900	118	85
Romania	1·1	3·2	2·7	3,140	118	90
SouthAfrica	2·4	3·9	3·2	2,740	112	78
Spain	0·9	3·4	3·0	2,600	106	81
Sweden	0·7	0·9	1·0	2,810	104	86
Switzerland	1·5	1·7	1·9	3,190	119	91
United States	1·5	2·0	1·6	3,330	126	106
USSR	1·5	3·9	3·0	3,280	131	101
United Kingdom	0·5	2·8	0·7	3,190	126	92
Yugoslavia	1·2	4·5	2·4	3,190	125	94
Developing Countries						
Afghanistan	1·9	1·7	2·2	1,970	81	58
Algeria	2·4	−0·8	3·4	1,730	72	46
Angola	1·8	2·7	3·0	2,000	85	42
Argentina	1·7	1·8	2·0	3,060	115	100
Bangladesh	3·57	1·67	··	1,840	80	40
Barbados	0·6	−0·1	··	··	··	··
Bolivia	2·3	5·0	2·7	1,900	79	46
Botswana	2·0	2·3	··	2,040	87	65
Brazil	3·0	4·4	4·0	2,620	110	65
Burma	2·2	2·4	3·3	2,210	102	50

Population, food supply and demand for food in individual countries

Table 1

	Population	Food production[1]	Demand for food[2, 3]	Energy supply[3, 4]		Protein supply[3, 4]
	Percentage rate of growth per year[5]			*Calories per caput per day*	*Percentage of requirement*[6]	*Grammes per caput per day*
Burundi	2·0	2·4	2·4	2,040	88	62
Cameroon	1·8	3·3	2·5	2,410	104	64
Central African Rep.	1·8	2·8	1·1	2,200	98	49
Chad	2·1	0·9	1·2	2,110	89	75
Chile	2·5	2·2	3·3	2,670	109	77
China	1·7	2·3	...	2,170	91	60
Colombia	3·3	3·1	3·9	2,200	95	51
Congo	1·9	2·2	3·7	2,260	102	44
Costa Rica	3·8	5·4	4·8	2,610	116	66
Cuba	2·2	1·1	2·0	2,700	117	63
Cyprus	1·1	5·4	2·3	2,670	108	66
Dahomey	2·3	1·5	0·1	2,260	98	56
Dominican Rep.	3·3	2·2	3·6	2,120	94	48
Ecuador	3·3	5·4	4·0	2,010	88	47
Egypt	2·6	3·4	3·8	2,500	100	69
El Salvador	3·0	3·6	4·1	1,930	84	52

Ethiopia	1·8	2·3	3·0	2,160	93	72
Gabon	0·6	3·6	2·4	2,220	95	57
Gambia	1·8	4·4	..	2,490	104	64
Ghana	2·9	3·9	3·2	2,320	101	49
Guatemala	3·0	4·1	4·2	2,130	97	59
Guinea	2·0	2·0	3·4	2,020	88	45
Guyana	3·0	2·5	3·6	2,390	105	58
Haiti	2·3	1·0	2·2	1,730	77	39
Honduras	3·3	4·0	4·2	2,140	94	56
India	2·1	2·4	3·0	2,070	94	52
Indonesia	2·5	2·0	2·6	1,790	83	38
Iran	2·8	3·3	5·4	2,300	96	60
Iraq	3·3	2·8	5·2	2,160	90	60
Ivory Coast	2·2	4·9	2·6	2,430	105	56
Jamaica	1·9	1·9	3·3	2,360	105	63
Jordan	3·2	1·8	6·6	2,430	99	65
Kenya	3·0	2·6	4·7	2,360	102	67
Khmer Rep.	2·8	3·5	4·3	2,430	109	55
Korea, Dem. Rep.	2·7	..	..	2,240	89	73
Korea, Rep.	2·7	4·8	4·7	2,520	107	68
Laos	2·4	3·7	3·7	2,110	95	49
Lebanon	2·8	5·0	3·1	2,280	92	63
Lesotho	1·6	0·5	..	..	..	..
Liberia	1·5	1·1	1·8	2,170	94	39
Libyan Arab Rep.	3·6	5·3	..	2,570	109	62
Madagascar	2·4	2·8	2·1	2,530	111	58
Malawi	2·5	4·7	3·7	2,210	95	63
Malaysia (West)	3·0	5·2	4·3	2,460	110	54

Population, food supply and demand for food in individual countries

Table 1

	Population	Food production[1]	Demand for food[2,3]	Energy supply:[3,4]		Protein supply:[3,4]
				Calories per caput per day	Percentage of requirement[6]	Grammes per caput per day
	Percentage rate of growth per year[5]					
Mali	2·1	1·6	4·3	2,060	88	64
Mauritania	2·0	2·4	3·0	1,970	85	68
Mauritius	2·6	1·3	3·0	2,360	104	48
Mexico	3·4	5·3	4·3	2,580	111	62
Mongolia	2·9	...	...	2,380	106	106
Morocco	3·0	2·8	3·3	2,220	92	62
Mozambique	1·7	2·7	3·2	2,050	88	41
Nepal	1·8	0·1	2·1	2,080	95	49
Nicaragua	3·0	4·9	3·9	2,450	109	71
Niger	2·8	4·1	2·2	2,080	89	74
Nigeria	2·4	2·0	3·1	2,270	96	63
Pakistan	3·0	3·0	4·2	2,160	93	56
Panama	3·2	4·3	4·8	2,580	112	61
Paraguay	3·1	2·6	3·4	2,740	119	73
Peru	2·9	2·9	3·9	2,320	99	60
Philippines	3·2	3·2	4·2	1,940	86	47
Rhodesia	3·4	3·9	4·1	2,660	111	76

Rwanda	2·6	1·8	1·9	1,960	84	58
Saudi Arabia	2·4	2·9	5·0	2,270	94	62
Senegal	2·2	3·3	1·2	2,370	100	65
Sierra Leone	2·0	2·4	3·9	2,280	99	51
Somalia	2·2	1·1	1·5	1,830	79	56
Sri Lanka	2·5	3·6	3·1	2,170	98	48
Sudan	2·9	4·3	3·9	2,160	92	63
Surinam	3·1	...	4·0	2,450	109	59
Syrian Arab Rep.	3·0	1·8	4·6	2,650	107	75
Tanzania	2·4	3·1	3·0	2,260	98	63
Thailand	3·1	5·3	4·6	2,560	115	56
Togo	2·3	5·4	2·4	2,330	101	56
Trinidad and Tobago	2·5	1·9	4·8	2,380	98	64
Tunisia	2·9	0·8	4·3	2,250	94	67
Turkey	2·7	3·0	3·8	3,250	129	91
Uganda	2·4	1·8	3·2	2,130	91	61
Upper Volta	1·8	4·7	1·2	1,710	72	59
Uruguay	1·3	0·8	1·2	2,880	108	100
Venezuela	3·5	6·1	4·0	2,430	98	63
Vietnam Dem. Rep.	2·7	...	...	2,350	114	53
Vietnam Rep. of	2·5	4·3	3·2	2,320	107	53
Yemen Arab Rep.	2·4	−0·2	3·9	2,040	84	61
Yemen, Dem. Rep.	2·4	1·6	−1·0	2,070	86	57
Zaïre	2·0	0·2	2·3	2,060	93	33
Zambia	2·9	4·3	4·8	2,590	112	68

Source: *Assessment of the World Food Situation: Present and Future*
Document E/CONF 65/3 United Nations World Food Conference

Notes to Table 1
1 Food component of crop and livestock production only (i.e. excluding fish production)
2 Calculated on basis of growth of population and *per caput* income, and estimates of income elasticity of farm value of demand in *FAO Commodity Projections 1970–1980*, Rome, 1971
3 Total food, including fish
4 1969–71 average
5 Exponential trend 1952–72
6 Revised standards of average requirements (physiological requirements plus 10 percent for waste at household level)
7 1962–72

(24 million metric tons on average). North American exports then did not exceed 5 million tons. The picture has now dramatically changed. In 1975 North American exports were over 90 million metric tons, and Australia and New Zealand exports were around 8 million tons. Of the developing countries only Argentina, Burma, Mexico, and Thailand were significant net exporters.

The developing countries' reliance on imports has serious implications on at least two critical aspects: firstly, the food security of these countries becomes greatly dependent on output in the few big producing countries and on the stocks they hold; secondly, imports constitute a serious strain on the balance of payments of these countries, and thus have wider effects on the overall process of development. For example, the value of cereal imports by the developing countries rose from $4,000 million in 1972/3 to an estimated

$9,000 million in 1973/4 as a result of the increase in food prices![1]

One is immediately led to examine the performance of the developing countries in food production: it increased by 3·1 percent in the decade 1952–62; in the following decade, 1962–72, the figure fell to 2·7 percent. In a sense the growth rates themselves are not menacing, for they are comparable to the growth rates of food production in the developed world, as Table 2 shows. But whilst in the developed countries the population growth fell from a yearly average of 1·3 percent in 1952–62 to 1·0 percent in 1962–72 the story is different for the developing world. There, population growth rate remained the same in both decades: an average of 2·4 percent. The result is that the growth rate of food output *per capita* fell from an average 0·7 percent annually (1952–62) to 0·3 percent (1962–72).

On the demand side, it must be noted that population growth is one determinant of demand; another is the effect of rising incomes. According to the Assessment Document of the World Food Situation prepared for the World Food Conference, in which disaggregated data were used, the increase in food production lagged behind population growth in 34 developing countries in the period 1952–72. When allowance is made for both the "population effect" and the "income effect",[2] the gap between demand and supply is significant for 53 out of the 86 developing countries for which data were available.[3]

---

1 See below.
2 On the assumption that the growth of population accounts for 70 percent of the total increase in demand and rising incomes account for the remaining 30 percent.
3 The figures of food output in the developing countries are aggregative figures. Within the category of developing countries, some fared better than others, with food production which recorded an average 5 percent growth rate or better. Some have also devised viable systems of rural institutions.

Rate of growth of food production in relation to population, world and main regions, 1952–62 and 1962–72
Table 2

	1952–62 Food production			1962–72 Food production		
	Popula-tion	Total	Per caput	Popula-tion	Total	Per caput
	Percent per year[1]					
Developed market economies[2]	1·2	2·5	1·3	1·0	2·4	1·4
Western Europe	0·8	2·9	2·1	0·8	2·2	1·4
North America	1·8	1·9	0·1	1·2	2·4	1·2
Oceania	2·2	3·1	0·9	2·0	2·7	0·7
Eastern Europe and USSR	1·5	4·5	3·0	1·0	3·5	2·5
All developed countries	1·3	3·1	1·8	1·0	2·7	1·7
Developing market economies[2]	2·4	3·1	0·7	2·5	2·7	0·2
Africa	2·2	2·2	—	2·5	2·7	0·2
Far East	2·3	3·1	0·8	2·5	2·7	0·2
Latin America	2·8	3·2	0·4	2·9	3·1	0·2
Near East	2·6	3·4	0·8	2·8	3·0	0·2
Asian centrally planned economies	1·8	3·2	1·4	1·9	2·6	0·7
All developing countries	2·4	3·1	0·7	2·4	2·7	0·3
World	2·0	3·1	1·1	1·9	2·7	0·8

Source: *Assessment of the World Food Situation: Present and Future*
Document E/CONF 65/3 United Nations World Food Conference

1 Trend rate of growth of food production, compound interest
2 Including countries in other regions not specified

This means that if the discrepancy continues the absolute difference between the two will increase to some 85 million tons by 1985, on conservative estimates.[1] We can thus visualize the problems attendant upon such a shortage, for the crisis is not one of supplementary products that are postponable or dispensable, but it concerns food. And, unlike other commodities, food is not postponable.

Such a world crisis has not actually occurred, but the signs are already there. Not long ago, the problem was one of how to get rid of accumulated food surpluses. In the early sixties, the Food and Agriculture Organization of the United Nations had programmes and plans for the disposal of surpluses. In those years, the United States Government paid farmers millions of dollars to leave land fallow. Hardly anyone then thought that in one decade food reserves would reach the present dangerously low levels. In mid-1973 the wheat stocks of the main exporting countries were at the lowest levels for 20 years.

For many of the developing countries the famine hazard is not something of the future, but is there now, at this moment. The danger is already there on the Western African coast. In the past six years, twenty-five million peasants and tribe members have continued to wait for the seasonal rains, the sources of their livelihood. When the heavens denied them they waited for the second season; the third; the fourth; the fifth. For six successive seasons rainfall was, at best, exiguous. Having necessarily consumed the seeds and the young animals, these people were unable to plant again

---

1 As the document on the *Assessment of the World Food Situation* has shown, these estimates are conservative for three reasons:
 a) no account is taken of the possible introduction of programmes to alleviate malnutrition;
 b) no allowance is made for changes towards a more egalitarian income distribution and consequently a rightward shift in the demand curve for food;
 c) no full account is taken of the increased commercial demand for certain foods brought about by the rapid urbanization which is occurring everywhere.

or to breed animals. So these flat tropical areas which once produced the strongest of African warriors were left barren and desolate, after a loss of life not less, according to *Time Magazine*, than a hundred thousand.

In other countries, we find hundreds of camps erected for the refugees from drought. In these camps there are hundreds of thousands of people, the food intake of each being one-third of what is needed. It is no wonder that in such conditions of malnutrition, diseases spread with alarming rapidity. Typhoid, dysentery, measles and colics are threatening the majority, and cholera threatened 15 thousand in the capital of Niger. In Chad, one of the members of a tribe which received a UN official carrying first aid requested him not to bring in any more medicines, on the grounds that death from diphtheria is faster and less painful than slow death caused by famine.

The New York Times of 24 September 1974 reported that in Bombay the penniless widow of a soldier wandered from crematorium to crematorium pleading to place the body of her child, dead of malnutrition, on the funeral pyre of a stranger. The police finally took the body to a morgue. There are also reports of mothers in Madhya Pradesh selling their children for food, and families in Assam struggling to subsist on grass, seeds and roots.

If we look at another country, Ethiopia, the danger is there, but it was belatedly discovered. The drought there caused disastrous results. The bureaucratic governors of the provinces tended to hide from the Central Government the dimensions of the disaster, so as not to lose favour with the Emperor.

In the spring of 1974, however, the loss of life increased so much that these bureaucrats were unable to hide the facts. Subsequent studies unveiled vast areas of famine, caused by a prolonged drought, not only in the eastern part of Ethiopia but also in the south and south-east. Particularly hit were the provinces of Wallo and Tigre. The famine affected at least 2 million farmers and herdsmen, killing uncounted thousands. The preliminary studies showed that in one region of the country the death toll of animals was 27 thousand cattle, 25 thousand sheep and 500 camels.

This sad picture gives a vivid view of the food problem which the world is now facing. But, at the same time, it does not yet give the whole story. It merely indicates the proportions of the crisis of which the world is becoming conscious, and for which we are trying to devise solutions.

The search for a solution will first and foremost depend on the success in diagnosing the causes of the danger as well as the danger itself. This is the subject of the next chapter.

Chapter
Two

The
Causes

This chapter does not purport to give an exhaustive analysis of the causes of the food crisis. It concentrates on a few key issues and explores the causes from the standpoint of different time periods: momentary, short-run, and long-run.

If the cause of the crisis had been restricted to events in 1972 and 1973, then the problem would have been transient, and I would not have used the word "crisis". By now it would have been over. The food crisis, however, still remains, and I can clearly foresee that when this book is published the readers may wish to add a few pages that further underline the gravity of the situation.

In the late sixties, there was widespread optimism about the world food situation. From 1967 to 1970, harvests in the developing countries were encouraging. Good weather and the "Green Revolution" technology in major food-deficit countries of the Far East combined to produce impressive results. In this region the increase in food production ranged between 4 and 6 percent in each of the four years. In India it was possible to build up government stocks of food grains to the unprecedented level of 9·5 million tons by mid-1972.

In 1971 bad weather was widespread in the developing countries, but their total production increased slightly. 1972 was again unfavourable for agriculture in both developing and developed countries.[1] As a result food production decreased, as the following index figures show:

Total food production   Table 3

	1961/5	1969	1970	1971	1972	1973
Developed countries	100	118	121	127	126	134
Developing countries	100	118	123	125	125	129

Source: *FAO*

While the index number for developing countries is 125 for 1972 (the same as for 1971), this is of course a weighted figure for all regions. It is significant, however, that in the Far East and the Asian centrally planned economies production diminished, while for the Near East region it rose. In 1973 the situation improved. However, on a *per capita* basis, food production of the world and of many major regions fell off. In many individual countries production declined steeply, resulting in serious emergencies. This was particularly so for the six Sahelian countries of West Africa: Chad, Mali, Mauretania, Niger, Senegal, and Upper Volta. The drought which affected these areas spread to northern Nigeria, northern Cameroon, and parts of Kenya and Tanzania.

In explanation of these phenomena and many others, such as torrential rains and floods in some parts of the USA and the Philippines and unusually warm winters in others, some weather experts are beginning to wonder whether the world is passing through a climatic change.[2] Scientists continue to disagree, however, on whether the recent weather observations are signals of a new trend or deviation from the present trends. Despite a considerable advance of knowledge in this field, much remains unknown, owing to the

---

1 In the USSR grain production fell by 7 percent and wheat production fell by 13 percent.
2 The word "climate" refers to an average of atmospheric conditions over a period of time.

fact that changes occur as a result of many interacting forces, such as radiant energy of the sun, the tilt of the earth, differentials in the absorption and retention of heat between land and water, and the presence of carbon dioxide in the air. The lack of meteorological records over long periods (hundreds of years) precludes the quantification of all these variables and the isolation of each.

But controversial as these views are, there are still some facts on which there is no disagreement. As Walter Orr Roberts mentioned in the course of the Rome Forum gathering on the world food problem (2–4 November 1974), it is a fact that in the middle of the higher latitudes, where most food in the northern hemisphere is grown, there has been a general cooling since 1950, amounting to 0·3° C approximately. Another point is that the period 1920 to 1960 was a remarkably favourable period, different from that which emerged after 1960 or that which prevailed between 1880 and 1920. And an important factor is that climatic variations are noticeably large in the marginal areas where the responsiveness of crops to weather is pronounced. Such areas are found in the semi-arid lands or in the fringe areas of deserts (for example, the Sahelian Zone), and in Asia where, as the population grows, there is increasingly more agriculture.

The weather is one factor which affects food production; another is the cost of agricultural inputs. I shall first concentrate on fertilizers, which constitute a key element in increasing agricultural production. In the past, additional food production in many developing countries was achieved by increasing the area under cultivation. Later, especially as limited availability of cultivable land is a constraint in some countries, a big proportion of increased production was achieved through increasing yields from given areas. It has been estimated that in the 1960's, 50 percent of the increase in grain production in developing countries was due to increases in cultivated areas; in the future, this proportion is expected to decrease further.

Although fertilizers are not the only input that could increase productivity, they nevertheless are of paramount importance. The availability and the low price of fertilizers in the sixties played a very positive role in the Green Revolution. By 1973 fertilizer prices had risen about threefold over the 1970 level. The economics of the fertilizer industry are intricate. To simplify, it is worth recording that both on the supply side and on the demand side there were factors which contributed to the rise in prices. In the 1960's fertilizer production increased in anticipation of large increases in demand. Important technological changes and the development of specialized transport facilities contributed to the growth of the industry. Demand was not big enough to clear the market, so excess capacity in fertilizer plants drove prices downward. In the early seventies, prices rose again as a result of sharply increased demand: for example, in the USA there was a considerable increase in the area cultivated in 1973/4, raising the consumption of nitrogenous fertilizers.

The high prices reduced purchases by developing countries, which sometimes paid exorbitant prices for "distress purchases". In addition to high prices, a scarcity of fertilizer has added another burden to the capacity of developing countries to increase food production, especially in the land-scarce countries of Asia.

On the demand side, the population variable is the most important factor in the food crisis. In 1974, total world population was estimated at 3,900 millions, a rise of 1,100 millions over 1957. World population is now increasing at about 70 millions annually, nearly twice the rate for 1950. By the year 2000, that is, twenty-five years from now, world population may rise to 7,000 or 8,000 millions!

As was shown in the previous chapter, the developing countries have succeeded in increasing their food production. But the population growth continues unabated. Indeed, the growth rate itself of population seems to be increasing, being highest in Latin America, followed by the Near East, the Far East, and Africa.

The disparity between the rate of population growth in developed countries (around 1·0 percent per annum) and in developing countries (around 2·4 percent per annum) has significant effects on the demand for food and on the whole process of development itself. Two-thirds of the world population now live in the developing countries. If these divergent trends in growth rates continue, three-quarters of the world population will live in the developing countries by 1985.

What demographers tell us about the mathematics of population is alarming. There is no reason to take the projections only to 1985 or 2000, somehow giving the impression that everything will end then. In 2025 the world population may reach the alarming figure of 16 billions. If one follows these projections, and if nothing is done to alleviate the food problem, then the frightening prospect of 1 million child deaths a month could come true.

Apart from the Malthusian-model analysis, there are other complex aspects of population. The demographic pattern is important. By way of example, take the case of India. According to figures released in October 1974, an estimated 42 percent of the present Indian population is below 15 years of age and only 15 percent above the age of 50. The International Institute of Population Studies in Bombay tells us that the fertile period of an average Indian is between 15 and 44. The implication of all this is that in the course of the next decade, when a significant number of the 42 percent enter the reproductive age, it is likely that population will grow at a faster rate than in the last decade.

The population problem is at the centre of the world's food problem. True, food production in the developing countries was considerable; in many countries the *proportion* of the population suffering from undernourishment has declined. But when we talk of hunger and malnutrition, proportions or percentages are not meaningful. The *absolute number* of hungry people has increased.

In short, the population problem is inextricably intertwined with the food problem. Perhaps it is well to realize that this is so not only concerning demand (more mouths to feed), but also concerning supply. This will be increasingly felt as population growth pushes food production to marginal lands; worse still, existing food-producing areas may deteriorate, as is already beginning to happen in the deforestation in the foothills of the Himalayas, with consequent increased flooding.

On the demand side, it is worth noting that the rise in incomes in the developed countries led to an increase in meat consumption and indirectly contributed to the rise in prices of grains. The reason for this lay in the increased use of grains for feeding livestock. The demand for food is inelastic with respect to the price.[1] It is obvious therefore that if supply does not keep pace with demand, prices tend to rise rapidly. But it is worth noting that when we are talking of price elasticity, other variables are assumed constant, for example, incomes. If incomes rise, these induce shifts of the demand curve to the right. In the example just cited of the increased consumption of meat, when more grains are fed to livestock, it may be said that by increasing meat consumption an increase in incomes has indirectly increased the demand for grains.

A picture of the price trends of wheat, rice and maize can be obtained from Table 4.

---

1 The price elasticity of demand measures the responsiveness of the quantity demanded to a change in price. Demand is inelastic with respect to price if an x percent change in price leads to less than x percent change in the quantity demanded. It is immediately obvious that towards the end of 1974 and early 1975, prices decreased from all-time peaks of 1973/4 as a result of record grain harvests in the USSR and the Far East.

Recent changes in export prices of selected
agricultural commodities
Table 4

	Wheat	Rice	Maize
	(US no. 2, hard winter, ordinary, f.o.b. Gulf)	(Thai, white 5 percent, f.o.b. Bangkok)	(Yellow no. 2, f.o.b. Gulf)
	US dollars per metric ton		
1971	62	129	58
1972	70	151	56
1973	139	368	98
1972 January	60	131	51
June	60	136	53
December	104	186	69
1973 January	108	179	79
June	106	205	102
December	199	521	113
1974 January	214	538	122
February	220	575	131
March	191	603	126
April	162	630	114
May	142	625	114
June	156	596	117
July	169	517	135

Source: *FAO Commodity Review and Outlook*

As has been mentioned earlier, trade now plays an important role in the food situation. This means that sudden large purchases could markedly affect world prices and stocks of food. For example, in 1972/3, the USSR, usually a net grain exporter, became the biggest world importer when it entered the market as a buyer for 20 million tons of wheat and 10 million tons of grain.[1] In the meantime, world trade in rice contracted slightly in 1972 and again in 1973 because of the scarcity of exportable supplies, thus increasing the demand for other cereals. The net effect of all this was not only a rapid rise in prices as a result of factors operating both on the demand side and the supply side, but also a reduction of stocks to dangerously low levels. The stock picture is given below:

Closing stocks in the main exporting countries
(in million tons)

Table 5

	1971/2	1972/3	1973/4
Wheat	48·8	29·0	20·7
Coarse grains	55·6	39·6	31·8
Rice	9·1	6·3	3·7

Source: *FAO Commodity Review and Outlook, 1973-4*, Rome, 1974

The underlying and immediate causes of the food crisis, which I have sketched here, throw further light on the magnitude of the problem. It is clear that if food production is to increase, many complex issues must be solved. The subject of food, however, has an urgency which goes beyond or should go beyond sheer economic calculations. Economists may begin to worry if the growth in *per capita* income is not satisfactory, but perhaps it is now time to concentrate on the concept of *per capita* food.

In a country where *per capita* food availabilities are barely sufficient for survival, the failure in the growth of food output implies the loss of lives of human beings. But in a situation

---

1 Towards the end of 1975, a grain sales agreement was concluded between the USA and the USSR. This agreement was tailored to respond to erratic and massive purchases by the USSR, with a view to lessening the resulting instability in the world market. The USSR will now import a specified amount of grain each year, regularly spaced out through the year.

where food availabilities are adequate, an economy may grow at fast or slow rates without endangering the lives of people.

On the supply side, a country may possess advanced technology, high-yielding seed varieties, as well as the necessary infrastructure; but there is an ever-present grim and ominous threat: the weather. Recent examples are the torrential rains and floods in the mid-western United States and the exceptionally warm winters in the eastern United States. The results of such weather aberrations on plantings and harvests can be very serious for food production.

The world is beginning to realize the global nature and the magnitude of the problem. Relief operations for shipments of food which took place in 1973 and 1974 have had some favourable effects. But as the experience of these years has shown, regular shipments of food and even air-relief operations by international organizations and other agencies do not constitute a solution. It is no longer a situation of a famine here and a famine there. If it is realized, by way of example, that one-half of the inhabitants of India are living in bare subsistence, then it must be a frightening situation if India faces a shortage in any one major food crop. The death toll in such an event would be of the order of tens of millions.

The Prime Minister of Sri Lanka, Mrs Bandaranaike, recently said: "Half of our foreign currency resources must now be earmarked to meet the rise in energy prices and the remaining half to meet the rise in fertilizer prices, and nothing in the end remains for development." Let us recall that two-thirds of the world's population is in developing countries. The food problem is ultimately linked with political stability. The problem of the availability of the loaf of bread led to coups d'état in Niger and in Thailand, threatened the rule of Haile Selassie in Ethiopia,[1] and forebodes civil wars.

---

[1] This text was written before the Emperor was overthrown.

If these facts are realized, the search for a successful solution becomes the only logical step. The solution this time cannot be in terms of relief operations. If, in the past, shortages were a few hundred tons, in future they could jump to 100 million tons. Likewise, a successful solution will not depend on the "generosity" of the advanced countries to the developing ones – the cost of which seems to be rising while its value in real terms tends to decline. And, finally, the solution cannot be one of bilateral agreements.

Chapter Three

The Rich and the Poor

Facts are constantly iterating an important question: has the era of cheap food ended? In order to remain optimistic, I must for the moment exclude another question: has the age of adequate food supplies gone?

The answer to my first question is both "Yes" and "No". If the process illustrated by the facts and figures of the last few years is to continue, then indeed the age of cheap food is gone. But, on the other hand, if these facts, frightening and threatening as they are, have drawn the attention of the world to the need for quick and positive action, then the answer could be "No".

The world food crisis has arisen despite the increase in the acreage of cultivated land and also despite the increased use of fertilizers. Until 1973, the USA paid its farmers large sums of money to leave millions of acres uncultivated. But despite the relaxation of supply-management policies the problem remains.

The advanced countries can perhaps mitigate the gravity of the problem by various humanitarian programmes of aid. But the developing countries must not depend on such programmes for ever, for ultimately they can be only palliatives and not cures. Such programmes are important now, but over the middle and the long run they can be only secondary.

A major factor is the rise in fertilizer prices, referred to in the previous chapter. In order to correctly gauge the effect of this, it is well to note that agriculturalists are agreed that the acreage of cultivated land does not any longer mean much; one needs also to know the amounts of fertilizers used, for productivity is not a function of acreage alone but also of inputs. It was the plentiful availability of fertilizers that helped to realize the Green Revolution in Mexico, and the high yields of wheat in India and in Latin America, as well as the "rice miracle" in south-west Asia.

In the last two years the exorbitant rise in fertilizer prices has imposed a heavy burden on the balance of payments position of some countries. The world price of one ton of fertilizer was $40 in 1971; in 1974 it reached $360, when it was available. The curious irony is that the farmer who can bear this heavy increase in the price of fertilizer is also the one whose need for it is relatively less. For example, the absolute difference in price between last year and this year will be an additional $4,000 million to be borne by American farmers, yet it is they who will reap proportionately less gain from fertilizers than others; for if the same amount of fertilizer were used by Indian, Sudanese or Somali farmers, or by others in developing countries, the proportionate increase in yield would be higher. The explanation of this is simple. For example, if one ton of fertilizers is used on virgin land the yield may reach ten tons of wheat; beyond that point, the extra returns from additional supplies of fertilizers tend to diminish. It is the rich countries where land gets its full share of fertilizers which can afford to buy them. The developing countries, which cannot afford the rise in prices, are those where land is short of fertilizers. The problem here does not rest on who will gain proportionately more, but on who can pay more.

If the reasoning is followed logically, the gravity of the crisis, to the world as a whole and not only to the developing countries, will be appreciated. Even the food-producing countries, which stand to gain financially from a world food shortage with the consequent rise in prices, realize that the perpetuation of the crisis, however lucrative it may be in the short run, is likely to lead to grave results in the long run. The increase in grain output required to keep pace with growth of population cannot be reached without increasing yields, which cannot take place without the application of fertilizers.

What is the solution then? As we pointed out earlier, the right approach is one which does not rely increasingly on aid but seeks to solve the problem from the roots.

Here is the contradiction which lies at the root of the problem: there is plenty of land which could be brought under cultivation, yet much of it lies in countries which cannot afford the fertilizer or inputs necessary to make it productive. At present only 93 million hectares of the 740 million hectares of arable land in developing countries are served by irrigation. Almost half of the irrigated area requires renovation and improvement. We can therefore readily identify a major factor in the crisis: the lack of the funds necessary for agricultural improvement in the developing countries.

The fertilizer crisis, accompanied by other natural events, brought the food problem to the world forum. Many explanations were given for the exorbitant rise in fertilizer prices, the least plausible of which attributes it to the rise in the price of oil, on the grounds that the main fertilizers come from natural gas. This explanation does not seem to have any force. The rise in fertilizer prices and food items generally began long before the rise in oil prices and before the energy crisis caught the attention of the world. This is clearly exemplified in the following quotation from the Shah of Iran:[1]

"In 1947 the posted price for a barrel of oil in the Persian Gulf was $2·17. Then it was brought down to $1·79, and that lasted until 1969. So there were 22 years of cheap fuel that made Europe what it is, that made Japan what it is. Then the price of wheat jumped 300 percent, vegetables the same, and sugar in the past six years increased by 16

---

1 *Time Magazine*, 1 April 1974, p. 25.

times. So we charged experts to study what prices we should put on oil. Do you know that from oil you have today 70,000 derivatives? When we empty our wells, then you will be denied what I call this noble product. It will take you $8 to extract your shale or tar sands. So I said, let us start with the bottom price of $7; that is the government intake. Suddenly everybody started to shout 'foul'. Why don't you use coal and shale for electricity or to heat houses, and keep this precious petrol for the petrochemicals for another 300 years to come?"

The truth which the Shah of Iran wanted to convey to the American readers is simple. He was essentially saying: You first raised the price of food exported to us. We therefore cannot but raise the price of the basic item that we export to you, which is oil. We have given you our oil at cheap prices for all these years and you could derive from it seventy thousand by-products, and you used it in achieving the progress that you have made. It is our turn now, as developing nations, to utilize the oil revenues for progress.

The rise in fertilizer prices was the result of a complex set of phenomena going back to 1962. Between 1962 and 1967, there was a big increase in fertilizer production capacity which occurred largely in anticipation of substantial increases in demand. On the supply side, the low energy costs and the advances in technology, which considerably lowered costs of production, were additional incentives to increase output. What in effect happened was that world demand did not grow as fast as anticipated, and there was a consequent decline in prices, which in the late 1960's were 50 percent lower than in 1964. Although the fertilizer industry was to some extent supported by increased aid shipments, closure of older and less efficient plants removed nearly 10 percent of existing capacity in the producing countries.

In the early seventies demand began to grow, and the effect of this on prices in the international market started to be felt in 1972. As prices rose, apprehensive buying followed which drove prices even higher. The price rise in

1972 was from 30 to 50 percent and by mid-1973 prices were double the 1971 level. In fact the increase in oil prices had a relatively small effect on the cost of producing nitrogenous fertilizers, and an even smaller effect on the cost of producing phosphatics.[1]

It is worth recording here one further point. The rise in oil prices did not accrue only to the oil-exporting countries, which are developing countries, but was equally lucrative to the oil companies themselves, which are of course Western companies. In the first three months of 1974, the profits of the big oil companies rose considerably compared with their profits in the same period in the previous year.

The effects of the rise in oil prices differ from one country to the other. For example, the USA's total imports account for 13·5 percent of its total consumption needs. Its imports from North Africa and the Middle East are only 2 percent. For Western Europe the situation is different: the corresponding figures are 59 percent and 47 percent. Japanese imports are 72·6 percent of total consumption while imports from North Africa and the Middle East are 57·4 percent.

There are a few countries which could partially absorb the rise in prices. Take the case of the USA. In April 1974, the US Secretary of State declared during the Special Session of the General Assembly that American oil imports in the year 1973/4 were paid for by the rise in the export prices of food, which in one year reached $9,000 million. Of this increase, $7,000 million was realized from the rise in American wheat prices (300 percent) and the rest from the sales of other products.

Turning to the developing countries, some of these could

---

1 The rise in phosphate rock prices nearly doubled the cost of P_2O_5 in diammonium phosphate and triple superphosphate, of which phosphate rock is a major component. Even then this increase in the cost of production, as well as the increases in freight rates that were beginning to occur, accounted for a relatively small part of the increase in delivered spot prices on the international market. This clearly shows that fertilizer was priced well above cost of production.

absorb the rise in food prices better than others. For example, Morocco, with its phosphate deposits; Malaya, possessing rubber; and Zambia and Zaire, with their vast resources of zinc, have succeeded in doubling the export prices.

Hence, one can understand the new terminology of the "fourth world". In addition to the Western World, the Eastern World, and the developing countries, there seems to be now a fourth grouping, closest to the poverty line. In this grouping there are the Indian sub-continent and tropical Africa as well as a large part of Latin America. With this fourth grouping "everything went wrong" as *The Economist* puts it; it will have to find some $6,000 million as a result of increases in the price of oil and food, excluding any expected rise in the price of industrial products.

Many solutions have been advanced to solve the food problem in the short run. The UN General Assembly suggested the formation of a world food reserve as an immediate remedy. If such a solution is to materialize, the only way of achieving it would be a reduction in consumption in the USA in one way or another. In this connection, it may be noted that the question of consumption in the developed countries is now becoming increasingly topical. So far, the term "malnutrition" has had the connotation of inadequate nutrition among the poor. Norway became the leading country to make the point that we wrongly assume that all is well with the pattern of nutrition in the developed world and that the rest should copy it. In November 1974, Norway became the first country to try to reduce the grain eaten by animals, in recognition of the fact that too much meat could be harmful to the health of the human consumer of the animal, as well as being wasteful of the world's limited grain supply.[1]

Similarly, in the case of fertilizers, reduction in consumption by America and other developed countries is the only solution in the short run in order to provide the millions

---

1 Compare the situation in the USA, where only about 10 percent of the corn crop is consumed by humans, and the rest goes to animals.

of tons of fertilizers of which the developing world is in urgent need.

It is worth noting that estimates show that the USA consumes annually over 2 million tons of fertilizers for non-agricultural purposes such as golf courses and lawns. The equivalent figure for Britain is 100 thousand tons (*The Times*, 29 October 1974). Over 90 percent of the world fertilizer production is in the developed countries; the developing countries produce only $7\frac{1}{2}$ percent of world output: around 6 million tons. In many instances, their production suffers from bottlenecks such as lack of spare parts for machinery, and idle capacity exists. They consume around 15 percent of world production. This means that 30 percent of the world's population consumes 85 percent of the fertilizer supply. In the longer run a massive expansion of fertilizer production capacity in the developing countries is necessary. In the short run there is urgent need for international assistance. The potential contribution of new lands and high-yielding seeds depends on the use of fertilizers.

As regards the oil-producing countries, these must be considered from two standpoints. Firstly, they are nearly all developing economies. Secondly, these countries are now in possession of resources which are increasing to such an extent that they could have an appreciable effect in saving the developing world from the calamities of the food crisis.

It is well to look first at the figures of oil revenues, and then to discuss their implications. The picture is given in Table 6. It is apparent that in 1974 the increase in oil revenues of the exporting countries as a result of the price differential was $67,000 million above the figure for 1973.

The big question is: where will all these funds, which have accrued rather suddenly, go to?

Naturally, it is the sovereign right of each country to utilize its resources in the manner it sees fit. However, here we are not discussing the issue from the standpoint of the right of each country. No. The issue at hand is how to tackle an urgent problem which threatens the world as a whole. Indeed, it is a problem which essentially concerns

Growth of Oil Revenues
of the Developing Oil-Exporting Countries
1972–4
Table 6

Country	Oil Revenues (in million US$)			Additional Revenue 1974 over 1973
	1972	1973	1974	
Arab Oil-Exporting Countries				
Algeria	680	1,095	3,700	2,605
Iraq	802	1,465	5,900	4,435
Kuwait	1,600	2,130	7,945	5,815
Libya	1,705	2,210	7,990	5,780
Oman	–	192	740	548
Qatar	247	360	1,425	1,065
Saudi Arabia	2,988	4,915	19,400	14,485
Syria	–	98	377	279
United Arab Emirates:				
Abu Dhabi	538	1,035	4,800	3,765
Dubai	–	169	658	489
Egypt	–	154	591	437
Bahrain	–	44	168	124
Total		13,867	53,694	39,827
Non-Arab Major Oil-Exporting Countries				
Indonesia	480	830	2,150	1,320
Iran	2,423	3,885	14,930	11,045
Nigeria	1,200	1,950	6,960	5,010
Venezuela	1,933	2,800	10,010	7,210
Total	6,036	9,465	34,050	24,585
Other Oil-Exporting Countries				
	–	1,040	3,997	2,957
Grand Total	–	24,372	91,741	67,369

35

the developing world to which these countries themselves belong.

* * *

There is now nearly unanimity on the establishment of a fund or funds to which the oil-producing countries would contribute. Such a fund would aim at surmounting the problems attendant upon the rise in prices of fertilizers, food and oil. The governments of these countries took the initiative in setting up or in suggesting the setting up of national and regional funds, for example The Kuwaiti Fund, The Arab Fund, The Saudi Fund, The Abu Dhabi Fund, The Islamic Bank, and others.

At another level, Iran suggested the setting-up of a new fund with a capital of $3,600 million, to be organized within the World Bank. Iran suggested that each of the twelve oil-producing countries pay $150 million, and this is to be matched by a similar sum from those Western countries which are traditional aid-giving nations.

Each of these proposals, some of which have actually started to function, has its specific objectives and its economic and social level. We are, however, facing a particular problem. Its magnitude, its complexity, and its threat are such that its solution cannot be subsumed under a hundred or two hundred projects which are within the framework of these funds. The food problem, as we have seen, affects six out of every ten individuals in the world. The food problem is a global problem. The failure to evolve a solution will not only result in the downfall of governments but will threaten the death from famine of tens of millions of people. So a comprehensive solution must be found to provide food today, and food for the 75 millions who will be added to the population every year.

Why don't we then set up a Fund for Agricultural Development in the developing countries, a fund the purpose of which would be not to provide palliatives but, from the start, to eradicate the ills, a fund which would utilize a fraction of the resources possessed by the richer countries in order to alleviate the plight of poverty?

The threat is famine. It is a threat which is big, is urgent, and is actually there. The problem is of resources to surmount this. But the resources are actually available. The issue is then, simply, why not use them to remove the threat?

It could be visualized that the administrators of these funds, national or regional, might be advised to give high priority to agricultural development. But this would constitute only part of the solution and not the whole solution. Financial resources alone are incapable of realizing economic and social development of the magnitude one has in mind. Funds alone remain merely liquid funds. Only when these are combined with technology, expertise, and the appropriate socio-economic organizational structure can one achieve, or come close to achieving, the desired results in terms of productivity. In the meantime, the problems of rural development in all developing countries are so great and so complex that a solution could not be realized without a specialized organ which would act as an international forum for gathering funds, expertise and technology.

In 1974, the developing countries advanced several proposals as to the constitution of the desired organ. During the Bangkok meeting of the Economic Council for Asia and the Far East, some countries suggested a Fertilizer Fund to meet the short and long-term needs of the developing countries. Other countries, meeting in the second session of Science and Technology, suggested the setting up of a Protein Fund.

These may be good ideas. But one thing is clearly absent, and that is a comprehensive look at agricultural development now, and twenty or thirty years hence. It is therefore clear that what the developing countries urgently need is a specialized organ which gives the dimensions of the problem and the appropriate solution.

It appears that the first step in the right direction is the setting up of an Agricultural Development Fund. Such a fund is not intended to replace any existing fund or project. It constitutes an additional project, aimed at solving a chronic and accelerating problem. Estimates show that in

37

1975, the additional burden on the balance of payments of developing countries, as a result of the oil price differentials, will amount to around $10,000 million.

* * *

We know that before 1939 the developing countries were net exporters of cereals. They exported 12 million tons of food grains annually. They are now net importers, their imports coming mostly from the USA and Canada.

We are told that the elimination of the tse-tse fly in Africa would release for cultivation an area larger than the total agricultural area of the USA!

Senator Humphrey of the USA says that if Americans eat one less hamburger a week, this would make some 10 million tons of grain available for food assistance.

We are told that if India's agriculture were so organized that her farmers, instead of producing a mere 1,010 lbs of food grains per acre, could be as productive as farmers in Egypt, who produce 3,515 lbs per acre, India's food grain surplus available for export would be double total worldwide trade in food grains in 1972, and national and international officials would then frantically be trying to prevent the collapse of the world food grain market rather than trying to prevent hunger! (*The Guardian*, 1 November 1974.)

And the fertilizer lavishly consumed on lawns and other non-agricultural luxuries could satisfy Asia's anticipated fertilizer deficit.

* * *

All these are "ifs". And many other "ifs" could be added. If, for example, the arable land on the whole planet were cultivated as efficiently as farms in Holland, the planet could feed 67 billion people, 17 times as many as are now alive.

But it is not "ifs" that we need now. The issue is precisely what must be done in the short run and the long run to solve the problem of food in the full realization that there is no substitute for food; that it cannot wait.

The international community must recognize that, for

millions of people, a small reduction in the consumption of food would replace acute malnutrition by death from starvation. It must recognize that the problem is immediate: that while we continue to argue and discuss, the numbers affected by malnutrition grow daily greater. It must recognize that all remedial measures are costly: storage of food is expensive;[1] so are the techniques for eliminating spoilage by pests; so is investment for future and greater production. But the cost of failure is to be counted not in millions of dollars but in millions of lives.

The responsibility falls on all, rich and poor alike. Without healthy bodies and healthy minds, all else is futile. Nothing demonstrates the interdependence of people so well as the common need of mankind for food.

---

1 Storing grain for a year costs 10 to 15 percent of its market value. Thus, at current prices, 80 million tons would cost between 1\frac{1}{2}$ billion and $2 billion for storage.

Chapter Four

The Myth of Industrialization

It is now an established historical fact that the developing countries, immediately upon achieving independence, embarked on industrialization. There was a craze for industrialization. It almost became a form of religion, and was deified, with a belief that one simply had to have it to achieve prosperity from the miracles it supposedly wrought. Its magic charm blinded the logical mind and agriculture appeared to be an unproductive activity in the face of the wonderful prospect of industrialization. In the course of this hysteria, the leaders of the developing countries got encouragement from everywhere: from economic advisers; from the developed countries; and from their own population. The advisers seemed to join in the fervour, and the industrialized world responded with enthusiasm to the desires of the developing countries.

The developed countries stood to gain from this state of affairs, on both economic and political grounds. On the economic side, the developed countries are the source of manufacturing equipment and of the expertise needed for industrialization. On the political front, the results are dramatic and thus useful to politicians currying public favour. This is different from investment in agriculture, where any projects are remote from the cities and engender none of the desired glamour; and where the gestation period is long and the results, after a protracted effort, are not so immediately apparent.

The demand for industrialization has continued for the last twenty years or so. The developing countries voluntarily followed this pattern and breathlessly gave industrial projects the highest priority. Political considerations weighed more heavily than economic considerations, and industrialization became synonymous with growth.

But is it? The developing countries failed to realize that what matters is not merely the setting-up of a new manufacturing industry, but essentially the cost structures and hence the profitability of such an industry. The results were in most instances disappointing: an output at prices relatively high in the world markets, and often inferior in quality. And this is what the fundamentals of economics teach. To industrialize for industrialization's sake is one thing, and to industrialize on sound economic grounds is another. The essential matter is the efficient use of resources: if a country is unable to generate a surplus in the first place and its agriculture remains underdeveloped, and if essential inputs for manufacturing are lacking, then how can gains be expected? Writers may philosophise about balanced growth, but in the end there seems to be a lot of truth in what Malthus wrote in his *Principles*, namely, that the development of industry cannot proceed without the development of agriculture. They are complementary. Agriculture provides the needed food and the raw materials, and is a source of demand for industrial products. Malthus is right about this. It is curious that we always tend to associate him with his theory of population, and forget his economics proper.

All this is not to say that there is no place for industrialization in the developing world. There is such a place. But there is a difference between chimerical schemes and those based on economic laws. In the pure economic sense capital accumulation and investment can be productive. But let us not forget that investment can be undertaken not only in manufacturing industries but also in agriculture.[1]

Another mistake perpetrated by the developing countries was a failure to take account of the rise in prices of food and of other agricultural products. The result of this was that the continual rise in food prices often nullified other gains. Egypt and India are two cases in point; and so was Yugoslavia, before realizing this imbalance and correcting it.

The industrialization in the developing countries in the last twenty years was, as we have said, encouraged by the advanced countries. This is evident in many ways. Until recently, virtually all the loans of the International Bank for Reconstruction and Development were directed to manufacturing industry in the developing countries, and not to agriculture. When Robert McNamara became head of the Bank, he expressed concern about this bias, especially in basically agricultural countries where agriculture was neglected, and he made a commendable attempt to redress the balance. After a few years, he succeeded in raising the share of agriculture in the Bank's loans, but only to a mere 12 percent. In an interview published in *Europa* (Vol. II, No. 6, March 1975) Mr McNamara clearly states that sound development of a society almost by definition means advancement of the welfare of all the peoples of a society. He draws attention to the marked "skewness" of income[2]

---

1 The industrial sector usually forms a tiny fraction of the Gross National Product of a developing country, where agriculture has a much bigger share. Accordingly, even if the growth rate of the industrial sector is significantly high, and that of agriculture low, the overall growth rate of the economy is likely to be low because such a growth rate is a weighted average of the growth rates of all the sectors of the economy. For example, if a country's agricultural sector forms 80 percent of GNP, and increases by 3 percent, while the industrial sector forms 20 percent and increases by 10 percent, the total increase is 4·4 percent.

2 In economics, the "skewness" of income is measured by what is called the Lorenz curve: on the vertical axis is plotted the income variable in percentages; on the horizontal axis, the population variable in percentages.

in developing countries. He says the World Bank deals with a hundred developing countries which have a population of 2,000 million people. Of these, roughly 40 percent are the "absolute poor", having a *per capita* income of $50 a year. Of the total of this group of 800 million poor, about 80 percent (or 640 million people) are in the rural areas. Hence the emphasis on rural development. Mr McNamara believes that the appropriate solution would be to raise the productivity and hence the incomes of the rural poor.

The developing countries realized that they were on the wrong road only when the world food problem became evident, the gravity of which has also been appreciated only very recently. The short-run phenomena of bad harvests, running-down of stocks, and famines were essentially the warning signal of something basically wrong in the relation of demand and supply of food in the post-war period.

As the previous chapters have shown, the rise in the prices of grains is unprecedented, as a result of the increasing demand. FAO estimates, referred to earlier, show that stocks are at their lowest levels, and its future expectations are for a worsening situation.

The food problem is now a world problem. Its effects on the developing world will, however, differ from one country to the other. In the Arab World the effects vary. There are essentially two groupings of countries: oil-producing and exporting countries, and countries which are not exporters of oil.

The first category includes Saudi Arabia, Libya, Kuwait, Iraq, Abu Dhabi, Qatar, Bahrain and Algeria. The revenues accruing to these countries enable them to face the rise in prices of food and of inputs for food production. However, they have problems in realizing a reasonable rate of growth in the economy and a reasonable standard of living. The difficulties can be summarized as follows:

(i) The lack of an agricultural and industrial base. Funds accruing to these countries have mostly been invested in speculative concerns in foreign countries.

(ii) Shortage of skills and backwardness of education.

43

There have been efforts to ameliorate this situation, but these countries will have to rely on foreign experts for a long time to come.

(iii) Many of the oil-producing countries have a very low density of population. Vast areas are still uninhabited. There has been an influx of foreigners to some of these countries.

All in all, despite shortages in food production and the need for the varied inputs for development, this group of countries will have a big surplus from oil revenues. They urgently need to invest these sums in order to develop as well as in order to ensure stability. Their present financial resources are not unlimited. Already availability of oil is beginning to diminish in Bahrain, and other countries, for example Kuwait, may run out of oil before the end of this century.

Countries which are not exporters of oil are Egypt, Sudan, Tunisia, Morocco, Lebanon, Jordan, Syria, North Yemen, South Yemen. Some of them are producers of oil to the point of self-sufficiency; others are net importers. Most of them import a big proportion of their food, as well as agricultural and industrial inputs. At the same time, there are no industrial and mineral resources to cover their needs. These countries are at the start of their development, and are moving at different paces. There are potentialities of land and water resources as well as human resources which could be utilized.

Clearly, these countries are in a less favourable position than the oil-exporting countries. The reasons are obvious: in addition to high food prices, the prices of inputs such as fertilizers, pesticides and agricultural machinery are also rising. These countries were also subjected to economic and political pressures more than the other group, with resulting increased obstacles in the development process, such as:

(i) A dearth of capital available for development in the agricultural, industrial or service sectors. These countries have a low *per capita* GNP, and urgently need to raise their standard of living.

(ii) The economies of some of these countries, in terms of manpower, resources and stocks, were adversely affected by wars or preparations for war for more than a quarter of a century. Others were beset by political and economic upheavals, and sometimes by natural disasters such as occurred in Morocco and Tunisia. It is therefore not surprising that some of the countries, hitherto creditor countries, became debtor countries.

(iii) Some of the countries face population pressures at the national level or at the level of regions within a nation. True, in many instances, there is a sizeable skilled manpower, but because of the dearth of investment in productive facilities there are economic pressures. After all, pressure is a relative term: population relative to the resources being utilized.

It is clear, then, that the economic situation of those countries now is difficult, and if present trends continue the situation will grow worse. Prices of imported food such as wheat, corn, rice, and sugar have soared. It is worth noting here that not all prices of agricultural goods rose in the same proportion. For example, whilst the price of wheat nearly quadrupled, that of cotton only doubled. The implications of this on the terms of trade of some countries are important: for example, for each ton of cotton Egypt exports, she can buy less and less wheat. Thus, worsening terms of trade may be experienced even for primary commodities, not simply between primary commodities and manufactured goods, as economists were prone to believe in the past. After all, primary commodities cover not only food but also agricultural raw materials and minerals, and the price trends of each of these may diverge even more than the price trends of primary commodities as a group and manufactures as another. Surely, from the point of view of any one developing country, what is relevant is the nature of the primary commodities that it exports, and the purchasing power of one unit of exports in terms of one unit of imports.

The demand for food by the Arab countries is not only

big, but is far from being satisfied. Egypt is a case in point. Its imports of wheat in 1974 reached the 500 million dollar mark. As the analysis of the terms of trade given in the previous paragraph shows, the implications for Egypt are strains not solely on her balance of payments, but also on overall development.

The sixties were characterized by a commendable growth rate. The seventies are clearly characterized by a new set of conditions. Many of the variables which hitherto could be considered constant are no longer so. The rise in food prices had serious implications not only because food is a basic need, but also by its effect on the whole development process. Let us not forget that there are millions in these countries who live from hand to mouth, or, if one may use another phrase, from land to mouth. Added to all this, there are political factors, and perhaps pressures, which must be taken into account.

In view of all this, the role of agriculture in the development process is becoming increasingly important and urgent. Developing agriculture is now the corner-stone of industrial development. Malthus, the economist, comes to life again.

Part Two

Problems and Solutions: The Arab World

Chapter
Five

Egypt and Sudan: Problems and Solutions

So far, one important fact emerges. The world is faced with a global food crisis. The situation in the developing countries indicates the size of the crisis and its explosiveness. The problem is certainly global, but perhaps if we reduce our scale we may see more clearly. I begin my discussion by looking at the Egyptian and the Sudanese economies.

The Egyptian Economy Despite the launching of industrial development in the past twenty years, agriculture has been the backbone of the economy. Agriculture accounts for 31 percent of the Gross National Product, and it employs one-half of the total working force. About 80 percent of the total export proceeds are accounted for by agricultural commodities. Through the years, agriculture has generated a surplus that spilled over to the rest of the economy. In Egypt the produce of the land satisfies consumption, except in wheat and vegetable oils, which are imported in large quantities. The production of other food items, such as rice, onions, and fruit, is in excess of domestic needs and these commodities are exported.

Production in the agricultural sector grew annually by 4·0 percent in the fifties and 3·5 percent in the sixties (until 1966). This meant that *per capita* output increased by 1·5 percent and 1·0 percent respectively, given that population growth is estimated at 2·5 percent annually. Since the late sixties, however, the performance in the agricultural sector has slowed down. On the supply side, availability of cultivable land is clearly a constraint. On the demand side, the rapid population growth has imposed strains on the country's *per capita* output.

The productivity of Egyptian agriculture varies from one crop to the other. In the case of wheat, beans, onions, lentils, cotton, corn, rice, and peanuts, yields are very high and in many instances compare favourably with yields in developed countries. In the case of vegetables and fruit, yields have been declining. Good weather conditions, diligent farming and intensive cropping have made Egyptian agriculture a profitable enterprise which has contributed to financing non-agricultural development activities, feeding the population, and absorbing large numbers of an increasing labour force.

The cost structure of agricultural production in Egypt depends on the price of seeds, fertilizers, fodder, services, and pesticides, in addition to labour costs. While other factors remain constant, a rise in the price of any of these inputs raises the cost of production. In 1970/1 the cost of the five factors totalled £E349 million, which is about a third of the value of total agricultural production of £E1,166 million. The distribution was as follows:

Table 7

	£ E million
Seeds	33·3
Fertilizers	103·0
Pesticides	13·0
Fodder	182·3
Fuel and maintenance	14·8
Obsolescence	2·6
Total	349·0

It is worth noting here that consumption of fertilizers in Egypt is proportionately higher than in other developing countries, and is likely to continue so owing to intensive cultivation, the disappearance of silt, and the shortage of organic fertilizers.

If the goal is an increase in agricultural production, this at once demands an increase in the inputs used, and a still more rapid increase is demanded if the goal is to increase productivity and not mere production. Agricultural machinery such as water pumps, tractors, and equipment for threshing and for pesticide-spraying must be used on a bigger scale. In addition to this, it is noticeable that the labour force in the rural sector is becoming a constraint owing to the spread of education and the drift to work in manufacturing industry and in services.

The overall increase in costs of production is a weighted average of the increase in the prices of the individual items. Fertilizers, accounting for a big percentage, weigh heavily in the total increase. The implications of the rise in fertilizer prices which we have seen are all too obvious.

The profitability of any agricultural enterprise is a function of the costs of production and the revenues. Revenues from individual crops do not necessarily rise by the same percentage as do costs. This is likely to have an adverse effect on the surplus from agricultural production. The issue is complicated by an additional factor. Not only have the prices of fertilizers, for example, risen markedly, but shortages in supply are now a major issue, especially since demand is increasing at unprecedented rates because demand is starting from a base far short of needs.

Despite the land constraint in Egypt and the heavy utilization of land, there is still room for progress. Recent changes in cropping patterns concentrating on high-value crops have contributed to growth in agriculture. Land reclamation projects and effective utilization of the water as a result of the High Dam could add to agricultural production. But despite Egypt's success in agricultural activity, the economic events of the present decade constitute serious problems to

the country. As a developing country, Egypt's import requirements for capital and inputs are understandably on
the increase. It follows that increased export receipts are
needed to finance such imports. We have seen that Egypt
is a big importer of wheat, the price of which has quadrupled
in the last few years at a time when the price of cotton
(by far the biggest agricultural commodity exported by
Egypt) has only doubled. The implication of this is a worsening in the terms of trade of the country. This simply means
that in real terms one unit of exports purchases fewer units
of imports. All this, followed logically, means that less and
less foreign exchange is available for investment and input,
much needed for overall development. But this is not the
whole story. We have seen that prices of inputs such as
fertilizers and pesticides have been rising alarmingly in the
world markets – yet another obstacle to the development
of agriculture. Hence, the situation is one where although
there is potential in terms of increasing yields and to a
small extent of increasing the cultivated area, the cost of
doing so is becoming an onerous burden on the economy.

Apart from what is happening in the international context
and its bearing on development, there is yet another problem
of a national character. It is a well-known fact that a developing economy, and Egypt is certainly no exception, subsidizes
food items as well as inputs such as fertilizers to the population. From the preceding analysis, the burden of this on
the Government's budget may be readily understood. The
Egyptian Government's subsidies to supply basic food as
well as fertilizers and pesticides to farmers have risen from
45 million Egyptian pounds in 1970/1 to 550 million pounds
in 1973/4. Despite these subsidies, food prices in Egypt have
risen faster than the prices of other goods.

Egypt may differ from other developing countries in that
the efficiency of its agricultural production is much higher.
Corn yields in Egypt are slightly more than three times
higher than in India. But Egypt serves as a good example
of the problems to be faced by all developing countries
in increasing agricultural production.

51

Since the Second World War about 80 percent of the increase in agricultural production in the developing countries has resulted from expanding the cultivated area. Fertilizers and advanced technology were the exception rather than the rule. The problem now is that countries will find themselves moving more and more in the direction of marginal lands, and the need for more costly investments and for additional use of fertilizers per unit of output will increase. In economic terms, this means that a lesser percentage of the cost of food will consist of land rent; that is, the share of non-land inputs will increase. This is happening at a time when costs of these inputs, particularly fertilizers, are at unprecedentedly high levels. We are in a situation where in many countries agricultural production is falling not only *per capita*, but also per acre.

The solution is not simply mechanistic or technical. There is much more to agricultural production than the traditional analysis of a production function linking inputs to output via some technical coefficients. The cold reality is that one cannot talk of increasing agricultural production without talking of rural development. The key element is the involvement of the people, particularly the poorest among the agricultural population, in the process of economic and social development. This implies a recognition that human resources are the most precious resources in the developing countries. Surely the process of utilizing these resources is not a simple one which can be resolved by decrees; it is a complex long-term process. Unless the farmer has a stake in the production process, there is really no solution.

Egypt may have succeeded in its land reforms because these were followed by efforts to assist the small farmers to organize themselves into viable institutions to improve the lands and to help in the control and management of water resources and in the application of technology. But these efforts must be sustained. And what of the tens of millions of farmers in the developing world who are still in the grip of village traders and money-lenders, where crops are mortgaged, and where exorbitant covert interest rates are charged? How can one expect a sub-subsistence farmer, under heavy debt and unable to feed his family, to use better seeds, to apply fertilizers, and to improve existing techniques of production?

The Sudanese Economy   In spite of its enormous area of agricultural land and its relatively small population of 16 million,[1] the Sudan has a food problem. The *per capita* calorie intake averages 1,440 calories. *Per capita* income is under $150 per annum. Before discussing this situation, I would like to go back into some historical facts about the country. Since the 1920's, cotton has been the most important cash crop. It is grown primarily for export. But durra (a kind of millet) has been the most important crop and has always occupied the biggest cultivated area in the country. It is mainly grown as a subsistence crop. The Sudan had a commendable record during the First World War and the Second World War as a supplier of food to other Middle Eastern countries. This implied that the country could feed itself and generate an exportable surplus. The Sudan is also one of the richest countries in the world in livestock.

In recent years, Sudanese net imports of food[2] items have represented some 14 percent of the total import bill. The Sudan is currently importing 200 thousand tons of wheat annually.

Agricultural productivity in the Sudan is very low. Durra yields are given in Table 8.

The average yields of durra not only fluctuate from year to year, but are noticeably low and do not seem to be rising through time.

Other major crops, such as sesame and groundnuts, are still grown in the predominantly traditional sector of agriculture and yields are very low indeed. Yields of groundnuts

---

1 Sudan has the lowest population density per square mile in the African continent.
2 That is, gross food imports minus gross food exports.

Area, Output and Average Yield of Durra in the Sudan

Table 8

Year	Area in million feddan	Output '000 tons	Yield metric ton per feddan
1963/4	3·3	1,350	0·411
1964/5	3·2	1,140	0·360
1965/6	3·2	1,095	0.342
1966/7	3·2	850	0·267
1967/8	4·7	1,980	0·421
1968/9	2·8	870	0·308
1969/70	4·4	1,495	0·344
1970/1	4·8	1,535	0·314
1971/2	4·7	2,080	0·443
1972/3*	3·9	1,325	0·337

Source: Sudan Government *Economy Survey* 1972
* Estimated
(Note: 1 feddan = 1·04 acres)

on average were 0·345 metric tons per feddan in the crop year 1963/4; the figure is 0·253 for 1971/2. Sesame yields are also falling, from 0·165 metric tons per feddan in 1964/5 to an estimated 0·121 in 1972/3. Yet yields of all these crops grown in the Sudan could be increased four-fold or five-fold.

The conclusion from all this is that the country has potential. The unsatisfactory performance so far may be ascribed to lack of funds, shortage of manpower and of agricultural inputs such as fertilizers. If that is the problem in the Sudan, it represents a crisis in the context of the present world food situation. We find 100 million acres in the Sudan ready for cultivation but totally unutilized at a time when food output at the global level is being outstripped by population growth. But whose fault is it? The Sudan, as a developing economy, has been expending vast efforts to expand agricultural production. Indeed, the Sudan is one of a few developing countries which are not neglecting agriculture. The Sudanese ten-year plan for the last decade and the five-year plan (1971–5) have accorded agriculture a high priority, and the Sudanese authorities have clearly demonstrated in practice that they know what rural development means. The Gezira Scheme as well as other schemes pursued lately testify to this.

But there are serious constraints. There is a financial constraint. A developing country such as the Sudan with a very low *per capita* income could not be expected to be in a position to provide adequate financing for its agricultural projects. When the Sudan resorted to the International Bank for Reconstruction and Development it succeeded in getting a loan of $15 million, repayable in five years at an interest rate of 9 percent per annum, for agricultural investment in an area of 600 thousand acres of rainlands. Such a loan, at an interest rate which is not concessional, allows the commencement of work on the land; more is needed, especially for operational costs. Experiments carried out in the Sudan itself have shown that new strains of seeds coupled with more sophisticated agricultural techniques could quadruple present yields. In the case of sesame, for example, it was shown that if the indehiscent sesame varieties are planted yields may quadruple. The country is aware of all this, but the issue is precisely the lack of capital and other resources to translate this into reality. Within its own limited capabilities, the Sudan is revising its agricultural priorities and trying to increase output. A good example of this is the introduction of wheat-growing in the Gezira Scheme in the early sixties. Output jumped from 37,000 tons in 1964 to 235,000 tons in 1974, while yields showed a remarkable increase from 0·41 to 0·74 metric tons per feddan.[1]

Egypt and the Sudan   It would appear, from what has been said in this chapter, that both Egypt and the Sudan

---

1 The fact that the Sudan is embarking on wheat production does not contradict the point made earlier about the imports of wheat from abroad. Demand for wheat in the Sudan has been increasing for at least two reasons: rising population, and switching from millet to wheat, owing to changes in taste and increases in income.

are facing obstacles in pursuing development. The exorbitant rise in world prices of food, fertilizers, machinery and other inputs is imposing strains on balance of payments which affect not only food output but the whole process of development itself. These factors are exogenous to the economy of each country, and neither Egypt nor Sudan can alter world prices of food or fertilizers· or machinery. But is there something that they can do together which will be mutually advantageous?

Let us look at the map giving the economic geography of the two countries together. In the south there are vast tracts of unused cultivable land (the Sudan). In the north (Egypt), cultivable land is limited. But whilst manpower and expertise are in short supply in the south, the opposite holds true for the north. There manpower and expertise are abundant and are under-utilized. In short, over the two countries there are enormous resources, natural and human, which are not economically used.[1] The result is very low productivity per unit of productive factor. Economic co-operation, in one form or another between the two countries, is likely to lead to economic gains. Broadly speaking, Sudan could provide the land and Egypt the expertise and the manpower. I am thinking of projects such as joint ventures where Egyptian farmers work on land not used by Sudanese farmers, for example in Gezira Managil and Khashm el Girba areas, where 40 percent of the land is not utilized. This could be worked out equitably by a sharing of costs and of profits.

As we move to the whole of the Arab region, the picture gets clearer and clearer.

---

1 I have in mind here the notion of economic efficiency, which means getting a maximum output from given inputs; or for a given output, using the minimum inputs.

Chapter Six

Food Production in the Arab World

We know that the Middle East is the ancient home of wheat, which was domesticated and developed as a cultivated crop from early times. We also know that barley and maize are other ancient crops of that region. In the world of today, when in practically all countries a lot of talking is taking place on cereal output, cereal deficits, cereal stocks, and so forth, it is well to look at the performance of the Arab countries in cereal production. Before doing so, however, I would like to make some comments about the general state of agriculture in the Arab countries.

Agriculture in the Arab countries today is a curious mixture of ancient and modern technology. The modern sector of agriculture forms a small proportion of the total agricultural sector. By and large, the present agriculture in the Arab countries has a low productivity, especially when one takes into consideration the favourable natural resources. Not only is productivity low, but there are vast unutilized tracts of cultivable land in Sudan, Iraq, and Syria. We have seen that in the Sudan over 100 million acres of cultivable land are not cropped. In Iraq only 4 million acres are cropped out of 17 million acres. Yet these figures represent an understatement of the facts, because with the use of modern technologies, investments could be undertaken to irrigate millions of acres more.

The pattern of trade among the Arab countries is worth

commenting on. Libya, Egypt, Jordan, Kuwait, and Saudi Arabia import wheat from Europe and the USA at a time when Tunisia, Algeria, and Morocco produce wheat in excess of domestic demand. Inter-Arab trade[1] within the whole region forms a small fraction of their total trade. The underdevelopment of agriculture is one factor which diminishes the volume of trade within the region. If agriculture were developed and agro-industries were established, considerable trade expansion within the region could take place.

There is another economic activity which is often lumped together with agriculture in the national accounts, and that is animal husbandry; I prefer the term "animal husbandry" to "livestock production", because the former has a wider context of meaning. Since my central theme is food production, it would be an omission not to refer to the important economic role of this activity in the Arab countries. A notion of the animal wealth of the region may be had from Table 9.

These figures can only be rough estimates.[2] In all the Arab countries data on livestock numbers are less accurate than data on agricultural crops. Nevertheless, these figures draw attention to this important resource. In Egypt, livestock products[3] account for about 30 percent of the gross value of agricultural production. The corresponding figure for Syria is 25 percent; for Jordan and Lebanon 30 percent. Thus, despite the fact that animal husbandry takes the second place as compared with crop production, its contribution is significant.

---

1 That is, in food items and all other commodities.
2 One reason for the unreliability of data, especially for sheep and goats, is the fact that a lot of the nomadic grazing of livestock is regional rather than national. Flock movements depend mainly on availability of grass and forage, which in turn depends on comparative rainfall in various areas.
3 Animals, milk, eggs, and poultry.

Livestock Numbers in '000 heads 1974

Table 9

Country	Cattle	Sheep	Goats	Camels
Algeria	950	8,100	2,400	190
Bahrain	4	4	8	1
Egypt	4,310*	2,080	1,280	110
Iraq	2,060	15,500	2,450	322
Jordan	40	670	360	9
Kuwait	7	100	80	6
Lebanon	84	227	330	1
Libya	121	3,200	1,110	120
Mauritania	1,800	2,800	1,900	710
Morocco	3,700	19,000	8,500	180
Qatar	6	39	45	9
Saudi Arabia	310	3,030	1,700	59
Sudan	14,000	11,900	8,600	2,620
Somalia	2,980	3,900	5,050	3,040
Syria	539	5,940	715	6
Tunisia	650	3,300	660	180
Yemen Arab Republic	1,250	3,500	8,100	61
Yemen People's Democratic Republic	99	230	915	40
(i) Total: Arab Countries	32,910	83,520	44,203	7,664
(ii) World Total	1,178,867	1,032,667	397,917	13,256
(i) expressed as percentage of (ii)	2·8	8·1	11·1	58·0

Source: Compiled and computed from *FAO Production Yearbook, 1974*
* Including an estimated 2,150 thousand buffaloes

The livestock economies of these countries remain under-developed. In most instances the animals obtain their feed from grazing on range lands, with little use of feed from harvested land. Cattle is concentrated in Sudan (mainly the southern part of the country), Egypt and Morocco, while the distribution of sheep and goats among the countries in the region is less uneven. Camels are clearly concentrated in Sudan and Somalia. Generally, cattle are native breeds, but some western breeds are found in all countries. Cattle perform a number of functions: they are used as work animals, for milk production, and in providing meat. Sheep are mostly range sheep. In Egypt, however, sheep are kept on irrigated farms; they are an important source of milk and meat. The production of wool from sheep is, however, very underdeveloped. Goats, like sheep, are an important source of milk, and they produce mohair and meat. In Egypt goats are kept on farms but in virtually all the other countries they are herded on natural pastures or ranges.

The predominantly nomadic character of livestock production in the Arab countries has effects on land use and on quality of livestock. Nomadic livestock graze on any forage available without control. One result of this is overgrazing or severe overgrazing. Overgrazing of range or grazing lands results in an actual decline of the forage production relative to the potential production from such lands. In addition to this, overgrazing means that the animals are poorly fed. This is more serious in the case of animals which travel long distances.

Another uneconomic use of the land is that of allowing weeds to grow on fallow lands when there is a grain-fallow rotation. The main reason for a grain-fallow rotation is to allow moisture to accumulate. If weeds grow, however, they absorb the moisture which would otherwise have been left for the grain. In view of the fact that the increments in moisture result in more than proportionate increments in grain, the losses can be considerable. In terms of quality, weeds are a poor livestock feed.

As Table 9 shows, the Arab World as a whole is rich in livestock production. But the livestock industry remains underdeveloped, and does have effects on protein supply.

I now turn to cereals, which understandably receive much attention in any analysis of the food problem. I shall analyse the cereal output in the Arab countries by looking at each of the main cereal crops. I shall then examine the situation for all the Arab countries in the context of world production.

Wheat Production in the Arab Countries in '000 metric tons
Table 10

Country	1961–5 (Average)	1972	1973	1974
Algeria	1,254	1,956	1,100	850
Egypt	1,459	1,618	1,838	1,984
Iraq	849	2,625	957	1,339
Jordan	180	211	50	225
Lebanon	64	64	55	60
Libya	37	42	67	70
Morocco	1,336	2,161	1,574	1,853
Saudi Arabia	129	150	150	175
Sudan	36	140	152	235
Syria	1,093	1,808	593	1,630
Tunisia	495	730	885	810
Yemen Arab Republic	21	54	50	71
Yemen Democratic Republic	15	15	15	15
(i) Total: Arab Countries	6,968	11,574	7,486	9,317
(ii) World Total	254,399	346,823	377,272	360,231
(i) expressed as percentage of (ii)	2·74	3·34	1·98	2·59

Source: Compiled and computed from data in *FAO Production Yearbook, 1974*

Wheat Table 10 gives a picture of production from 1961 until 1974. Out of the thirteen Arab countries given in the table,[1] it is immediately apparent that six countries stand out as substantial producers. These are Egypt, Morocco, Syria, Algeria, Iraq, and Tunisia. Another point worth noting is that, with the exception of Egypt, the remaining five substantial wheat producing countries exhibit very high instability in output. In Iraq and Syria total output in better years is often three times total output in poorer

1 It will be noticed that in this and subsequent tables only thirteen Arab countries are included. The reason for this is that other Arab countries, for example United Arab Emirates, Qatar, Oman, Bahrain, and Kuwait are not cereal producing countries. The tables therefore give a complete picture of cereal production by all the Arab countries.

years. The reason for this is that wheat production in those countries depends on rainfall. Since the Second World War, wheat production in Syria and Iraq has expanded to sub-marginal lands where annual rainfall is inadequate or variable or both. In Egypt, wheat is artificially irrigated, and output has been growing without fluctuations. Yields in Egypt are much higher than in any of the other countries (see Appendix 3). Indeed wheat yields in Egypt are higher than the average yields in the USA, where the plant grows in rainfed areas. By contrast, in Egypt all wheat is irrigated wheat.

Barley Production in the Arab Countries in '000 metric tons

Table 11

Country	1961–5 (Average)	1972	1973	1974
Algeria	476	720	400	450
Egypt	137	109	97	99
Iraq	851	980	462	532
Jordan	62	34	6	35
Lebanon	13	8	7	8
Libya	87	116	205	100
Morocco	1,316	2,468	1,257	2,389
Saudi Arabia	10	20	18	22
Sudan	–	–	–	–
Syria	649	710	102	655
Tunisia	145	236	210	300
Yemen Arab Republic	141	178	150	230
Yemen Democratic Republic	3	4	4	4
(i) Total: Arab Countries	3,890	5,583	2,918	4,824
(ii) World Total	99,686	153,309	169,245	170,858
(i) expressed as percentage of (ii)	3·90	3·64	1·72	2·82

Source: ibid.

Barley As Table 11 shows, the biggest producer of barley among the Arab countries is Morocco. Yield fluctuations (see Appendix 4) are less pronounced than for wheat, for although in all of these countries, except Egypt and Iraq, barley is grown on rainfed lands, it is more tolerant of salts in the soil than is wheat. Also, it requires less moisture and hence less rainfall. These factors have rendered barley production less erratic than wheat production.

The Appendix table of barley yields also shows that yields per hectare are highest in Egypt. Indeed, barley yields in Egypt have been rising remarkably for the last fifty years. But as part of deliberate Egyptian policy, the area has been decreasing. This is a result of the comparative advantage of growing barley as against other crops in Egypt where land is a constraint. On the demand side, the market has not been expanding fast.

Maize Table 12 shows that of all the Arab countries, Egypt is by far the biggest producer, supplying some 80 percent of the total output of the region. Next is Morocco. Maize is an important crop in Egypt, where it fits into the rotation as a late summer crop. The rise in Egyptian output is due to successively increasing yields (see Appendix 5), owing to the introduction of hybrid varieties since 1960. In Morocco, the second largest producer, yields are one-fourth or one-fifth of those in Egypt, owing to the unreliability of moisture supply as a result of erratic rainfall.

Table 12 also draws attention to the steadily increasing output of maize in the Yemen Arab Republic and the Yemen Democratic Republic. It is also worth noting that yields in these countries are two or more times higher than in Morocco (see Appendix 5).

Rice As Table 13 shows, over 90 percent of the rice grown in the Arab countries is produced in Egypt. Nearly all the rest is grown in Iraq. Rice yields per acre in Egypt are the second highest in the world, exceeded only by Australia.

Maize Production in the Arab Countries in '000 metric tons

Table 12

Country	1961–5 (Average)	1972	1973	1974
Algeria	4	5	4	5
Egypt	1,913	2,421	2,508	2,550
Iraq	2	18	19	19
Jordan	–	–	–	–
Lebanon	12	1	1	1
Libya	2	1	2	2
Morocco	352	368	217	389
Saudi Arabia	2	5	6	6
Sudan	17	11	18	20
Syria	7	15	15	19
Tunisia	–	–	–	–
Yemen Arab Republic	39	42	51	150
Yemen Democratic Republic	9	80	75	84
(i) Total: Arab countries	2,359	2,967	2,916	3,245
(ii) World Total	216,381	305,388	310,391	292,990
(i) expressed as percentage of (ii)	1·09	0·97	0·94	1·1

Source: ibid.

In Iraq, yields are about one-half of those of Egypt (see Appendix 6).

There are two other cereal crops produced in the Arab World: millet and sorghum. Egypt is the biggest producer of millet (see Table 14) followed by Sudan. These two countries together produce some 90 percent of the total output in the region. Sorghum (see Table 15) is produced mainly by Sudan, followed by the Yemen Arab Republic, on rainfed lands, and is subject to high instability in output.

So far, the discussion has been confined to individual cereal crops in the Arab countries. We now turn to examine the performance of the Arab countries in the context of world production.

Rice Production in the Arab Countries in '000 metric tons

Table 13

Country	1961–5 (Average)	1972	1973	1974
Algeria	7	5	6	5
Egypt	1,845	2,507	2,274	2,400
Iraq	138	268	157	275
Jordan	–	–	–	–
Lebanon	–	–	–	–
Libya	–	–	–	–
Morocco	20	14	14	18
Saudi Arabia	–	–	–	–
Sudan	1	5	6	7
Syria	–	–	–	–
Tunisia	–	–	–	–
Yemen Arab Republic	–	–	–	–
Yemen Democrat. Republic	–	–	–	–
(i) Total: Arab Countries	2,011	2,799	2,457	2,705
(ii) World Total	253,180	295,608	324,468	323,201
(i) expressed as percentage of (ii)	0·79	0·95	0·76	0·84

Source: ibid.

For each of the six cereal crops, the output of all the Arab countries is expressed in the tables as a percentage of total world output. The figures are self-explanatory. Wheat output by the Arab countries ranges from nearly 2 percent to 3·3 percent of total world output, and seems to be unstable, with no clear-cut trend. Nevertheless, it is worth noting that output in each of 1973 and 1974 forms a smaller percentage of total world output than in 1972. In the case of barley, the proportion of output by the Arab countries to world output seems to follow a declining trend. For maize the trend is fairly constant, and similarly for rice, if the year 1972 is omitted. It is interesting to note that rice output by these countries is less than 1 percent

Millet Production in the Arab Countries in '000 metric tons
Table 14

Country	1961–5 (Average)	1972	1973	1974
Egypt	723	831	853	875
Sudan	303	355	268	470
Other Arab Countries	181	241	221	260
(i) Total: Arab Countries	1,207	1,427	1,342	1,605
(ii) World Total	38,159	42,489	48,067	46,215
(i) expressed as percentage of (ii)	3·16	3·36	2·79	3·47

Source: ibid.

Sorghum Production in the Arab Countries in '000 metric tons
Table 15

Country	1961–5 (Average)	1972	1973	1974
Sudan	1,256	1,300	1,625	1,795
Yemen Arab Republic	933	1,020	1,000	720
Other Arab Countries	262	270	245	258
(i) Total: Arab Countries	2,451	2,590	2,870	2,773
(ii) World Total	35,779	45,583	53,206	46,908
(i) expressed as percentage of (ii)	6·85	5·68	5·39	5·92

Source: ibid.

of total output. With sorghum the situation is different, where output by the Arab countries is more than 5 percent of total world output. For millet, output by the Arab countries ranges between 2·8 percent and 3·5 percent of total world output.

If a more aggregative approach is followed, where *total*

cereal output by all the Arab countries is examined against *total* world cereal output, which includes other cereal crops besides the six already discussed,[1] the following picture emerges.

Table 16

	1961–5 (Average)	1972	1973	1974
(i) Total world cereal output (in '000 metric tons)	988,538	1,278,744	1,376,017	1,333,864
(ii) Total cereal output by Arab countries (in '000 metric tons)	18,886	26,940	19,889	24,469
(ii) expressed as percentage of (i)	1·91	2·11	1·45	1·83

Source: Computed from data in *FAO Production Yearbook, 1974*

Thus in 1974 all the Arab countries produced 1·8 percent of world cereal output. Population of all the Arab countries was estimated at 144·9 million[2] in 1974, whilst world population is about 4,000 million. If the population of the Arab countries is expressed as a percentage of total world population, the figure is 3·6 percent. It is curious to note that 1·8 percent is exactly one-half of 3·6 percent. This implies that if the Arab countries double their cereal output, one would have a situation where 3·6 percent of world population produces exactly 3·6 percent of world cereal output.

Can the Arab countries do this? From the preceding analysis it is clear that the potential is vast. Indeed, there is nothing sacrosanct about talking of just doubling their cereal output. After all, world output of cereals still falls

1 For example, rye, oats, mixed grain, and buckwheat.
2 See Appendix 2.

far short of adequately feeding the world's population, and we have seen in Chapter One that hundreds of millions of people are undernourished.

With additional investment in agriculture, with better marketing facilities, and with regional economic co-operation, the Arab countries can once again resume their earlier dominant position in the world grain market, not only producing to meet their own needs adequately, but becoming net exporters.

Chapter Seven

The Case for Regional Co-operation

The discussion in the previous chapters underlines the gravity of the food crisis at the world level. At the regional level, having chosen the Arab countries as an illustration of both the problem and the cure, I would like in this chapter to outline my thinking as to what the Arab countries could do to foster agricultural development. The picture is one where the arable land frontier in the region as a whole is still very far from being reached, implying that one does not have to resort to the cultivation of poor marginal lands in order to increase agricultural production. This means that, other things remaining constant, further expansion of agricultural production need not take place at increasing costs.[1]

But other things need not remain constant. The previous chapter has drawn attention to the very low agricultural productivity in most of the Arab countries. With the use of inputs such as fertilizers and pesticides, agricultural productivity could be increased enormously. At present the use of fertilizers is high in Lebanon and Egypt, but is exceptionally low in the other countries.

---

1 I use this term in its strictly technical meaning in economic theory. The law of increasing costs refers to a situation where one incremental unit of output costs more to produce than the cost of each of the previous units of output.

The three basic requisites of agricultural development are: land and labour availability; capital; technology and expertise. My thesis is that *enormous possibilities exist for regional co-operation in the supply of inputs to agriculture.*

If agricultural development is now a necessity, then plans must be made, in the context of an integrated approach for the region as a whole, to foster agricultural development. The time has come for these countries to make headway. Planning for the future calls for an appraisal of present planning and performance. So far, whatever planning has been done did not take account of the resources of the region as a whole. It was carried out by individual countries, within the confines of national resources, and very often those individual plans did not give priority to agriculture. In Egypt and other Arab countries much attention was given to industrialization. Industrialization in these countries was motivated by the desire to reduce imports, rather than to provide exports. The results have often been disappointing, not serving the cause of economic development. These have been inefficiency, waste of capacity, high unit costs as a result of gearing production to a market far smaller than the optimum. These industries have aimed at the domestic markets and many attempts at exporting have failed because of lack of competitiveness in quantity and quality. Import duties have been levied on foreign products, a practice defensible for the protection of infant industries, but not defensible if it is continued indefinitely. The argument for protection assumes that eventually the industry will cease to be infant and will be able to compete in the world's markets, when the tariffs would be lifted. But this has not happened.

Another issue worth examining is the volume of investment. Table 17 gives figures of planned investments in all sectors for development.

These planned figures were prepared in most cases in the late sixties. But since then important changes have been taking place in the world's economic situation. As we have seen, the prices of agricultural inputs and of food, both of them imported in large quantities by the Arab World, have recently risen to unprecedented levels. In these changed circumstances, a drastically different approach is called for. After all, this appears a logical conclusion when one realizes that the variables on which calculations were based are themselves changing. It seems to me that planning should no longer be for individual countries, but for the region as a whole.

It is clearly advantageous to the countries of the region to embark on economic integration. The purpose of economic integration is to utilize the resources of the region for the benefit, in real terms, of all members. Economic integration is now a common feature in present-day life. There are the EEC, EFTA, Comecon; and among the developing countries we have the East African Common Market, the Central American Common Market, and the Caribbean Free Trade Area. In fact, when economic integration was

Table 17

Country	Planned annual investment in million US $ *
Saudi Arabia	817·0
Iraq	600·0
Kuwait	468·0
Lebanon	369·0
Libya	1,542·0
Morocco	487·0
People's Democratic Republic of Yemen	24·0
Tunisia	567·0
Algeria	1,372·0
Jordan	167·0
Syria	425·0
Egypt	1,416·0

Source: National Statistics of each country.

* The plan period and the year in which the plan commences vary from one country to the other. In most instances it is the first half of the seventies which is covered.

first expounded in economic theory books[1] the analysis was mainly concerned with developed economies. In the sixties, economists began to analyse the effects of integration among developing countries and pointed out the dynamic gains[2] that one could expect from such integration. The benefits accrue from a decrease of unit costs when production is geared to a larger market. Integration minimizes the hazards of relying mainly on one crop for exports, such as wide fluctuations in prices and in proceeds of exports. In the context of the world economy, integration strengthens the bargaining position of the region vis-à-vis the rest of the world.

Generally speaking, a plan for economic integration must rest on the following basic points: (i) reduction or removal of trade barriers; (ii) free movement of capital; (iii) free movement of labour; (iv) unification of monetary and financial regulations; and (v) co-ordination of production policies.

Let me be more specific. I am not venturing in this book to write a treatise on economic integration in the Arab countries, because I do not want to depart from my central theme: food production. In the Arab World, economic conditions for integration are propitious: the Arab World has billions of dollars, millions of acres of cultivable land, millions of skilled labourers, massive water resources, and a climate in many ways favourable. Its position on the earth's surface is advantageous and its area is vast, extending 6,000 kilometres from west to east, 3,500 kilometres from north to south.

But the Arab World still remains a heavy importer of food, and many countries in the region face exorbitantly high food import bills since 1973. Protein deficiency may not be as pronounced in the Arab countries as it is in India,

---

1 See J. Viner, *The Theory of Customs Unions*, 1951.
2 Economic development is an activity happening through time. If economic integration takes place between countries and has net beneficial effects on the development process itself, it is said to have *dynamic* gains in contradistinction to *static* gains.

Pakistan or Bangladesh in Asia, or north-eastern Brazil in Latin America, or the Sahelian countries in Africa. But undernourishment is still there in the Arab World. Of course in its advanced forms undernutrition is recognizable in swelling bodies, peeling skins, and other symptoms. But the real danger is that even when it is not outwardly recognizable it impairs both physical and mental health and its effects are unfortunately irreversible when it occurs in early years.

The theory of economic integration or customs unions appears simple when formulated in economic analysis. In practice, there are difficulties, political, social, and economic, that stand in the way. The experience of the EEC countries is illuminating in this respect. The Brussels convention is now nearly two decades old, but progress in the Market countries in achieving gradual integration and in smoothing the attendant and inevitable readjustments has been steady but unspectacular.

Recent experience suggests that the establishment of regional companies is a practicable solution. In current economic literature, these have now come to be called "limited integration attempts". They represent a realistic approach to full economic integration, which, commendable as it is on economic grounds, cannot be expected to take place quickly. It is possible to revolutionize agriculture in the Arab region by beginning to undertake joint ventures which would be of mutual advantage. Existing projects could be used as a starting-point. Examples are: the sugar cane projects in Egypt, Sudan and Syria; the projects to expand cultivation of wheat in Iraq, Syria and Sudan; projects for expanding vegetable cultivation in Egypt and Algeria; projects for developing animal resources in Sudan, Libya and Algeria; poultry projects in Kuwait, the Gulf States and Egypt; forestry and stock-raising projects in Syria, Sudan and Algeria; and many others.

There are scores of such projects which could be undertaken as joint ventures. They would have the advantage of being large units with big capital investment and with considerable technological expertise, so that the competitive

73

position of these countries in the world markets would be strengthened. They could utilize human resources flexibly and could disseminate expertise. Production units could be set up in the countries in accordance with the principle of comparative costs.

A point which, it seems to me, is of crucial significance is the lack of linkages between crop production and animal husbandry in the Arab countries. The previous chapter has shown that the Arab countries are rich in animal resources. At the same time, I referred earlier to the undernourishment that affects sections of the populations of these countries. The development of the livestock industry in these countries is essential for meeting the growing demand for meat, which is an item generally characterized by a high income elasticity of demand.[1] Libya, Saudi Arabia and other Gulf Arab countries, hitherto exporters of meat, are now net importers.

How can co-operation be achieved in the context of the economic integration I have been talking of? Take the case of the Sudan. I shall outline what could be done to develop livestock in conjunction with other crops. Conceivably a joint venture could be set up in a given area, say in western Sudan, where fodder is grown for sheep, as well as oilseeds such as groundnuts. The sheep could be fed on the fodder as well as on the groundnut haulms. Crushing mills might be installed, to derive a useful top-grade edible vegetable oil from the oilseed.[2] Part of the resulting residue, the oilcake, could be used for fattening the sheep, and the rest could be exported. Modern slaughterhouses could be built, and refrigeration facilities installed. The meat could be exported. It is fitting to record here that thanks to the development of refrigeration there was a revolution in New Zealand's meat industry.

1 This means that, *ceteris paribus*, a rise in income of x percent, results in a rise in the quantity demanded of *more than* x percent.

2 Groundnut oil is a high grade edible oil. It may be mentioned here that since the Second World War the supply of vegetable oils has been limited, so much so that other oils, for example, linseed, once considered of the industrial oil variety, are now refined for use as edible oils.

Apart from the oilseed industry, other industries could arise, such as tanning, manufacture of glues, fertilizers from blood, bone meal (which could be used to feed poultry), and others. It does seem that such projects are exceptionally lucrative. The results may be the export of good quality meat likely to fetch high prices; the establishment of some agro-industries which generate employment and contribute to the training of labourers, and industries directly connected with livestock such as good tanning and proper flaying of skins. (It has been estimated that if, in the Sudan, tanning and flaying of skins were improved, prices of hides and skins could be doubled.) The markets for the end products are strong markets where prices follow healthy trends, unlike many other primary commodities which suffer from worsening terms of trade. The price trends for red meat, unlike those for white meat such as poultry, are rising, and the market for hides and skins is growing, while the market for oilcake has been soaring. All these profitable outlets, along with other beneficial effects,[1] can only make such projects very attractive.

As we have seen in the context of the Arab World, the real constraint on such activities is not the lack of availability of land but the dearth of investment. One country alone may not be able to undertake such investments, but economic co-operation with other countries in the region – a joint venture – provides the solution. By the same reasoning, investments in infrastructure and irrigation activities can forestall the vagaries of nature attendant upon rainfed agriculture, and contribute to raising output and averting instabilities in production and farm incomes. We always talk of the vicious circle of poverty; but the circle can be rendered benign if investments are undertaken. The results will be an increase not solely in the direct supply of food but also in profitable activities providing resources for future investments. Accordingly, a building-up process sets in.

1 For example, when modern crushing mills are established, the output of oil for a given tonnage of groundnuts exceeds the output achievable by crude traditional methods.

What attempts have so far been made by the Arab countries in the direction of economic integration? Early in the 1950's the Arab countries embarked upon fostering economic co-operation. But over twenty years later the record is one which at best is characterized by an increase in bilateral trade agreements. Despite the decision on 1 January 1965 to establish an Arab Common Market, no progress was achieved. All that happened was a series of bilateral trade agreements, which were concluded for short periods of time, with the lists of goods benefiting from preferential treatment continuously changing. But such agreements were never, and could hardly have been, conducive to a structural transformation in the productive structures of the respective countries. Very often these trade agreements were carried out on an *ad hoc* basis to dispose of surpluses. Unless the countries concerned are aware of the net benefits to be reaped from economic co-operation, and unless a more appropriate institutional structure is set up, integration cannot take place.

Chapter Eight

Oil Monies

Oil is now a topical subject that preoccupies the mind of virtually every country. Europe and Japan are almost totally dependent on Middle East oil. The USA is much less dependent on Middle East oil supplies, which account for only 6 percent of its total imports, but it will increasingly depend on them in the future, unless alternative energy sources are discovered.

The recent oil crisis has shown the importance of this raw material in the world economy. For many years its price was very low relative to other raw materials, and it played a crucial role in the industrial expansion of Europe, USA and Japan. Between 1960 and 1970 demand grew at 10 percent a year. There is no reason to suppose that demand is likely to decrease in the foreseeable future. Nuclear energy is still in the embryo stage, and even if technical, managerial and environmental problems are overcome, it is no secret that it takes seven to eight years to construct a plant. Oil from other sources (seas or rocks) can be extracted at high prices.

Not only have oil revenues increased in the Arab World, but there are basic changes taking place in the institutional structure of the oil companies. In 1972 the Gulf States pressed for participation in the companies' capital and got a 25 percent share. In Kuwait, Saudi Arabia, Abu Dhabi, and Qatar this percentage is expected to reach 51 in 1982.

It may happen that events will move faster. Participation not only results in increased revenues for these countries, but the degree of control they now possess enables them to participate actively in investment, production and management.

In the foreseeable future, the oil-exporting countries will continue to be in a powerful position. Whilst oil revenues for the OPEC countries were $5,000 million in 1970, they reached $92,000 million in 1974. For the whole period from 1973 to 1980, the Arab oil-exporting countries are likely to get $550,000 million in oil revenues at 1974 prices. Whatever government expenditure takes place, the surplus is going to be enormous. The economies of these countries, in their present structures, cannot absorb such sums. It is for this reason that these countries should think now and immediately of ways and means of using these resources for their own development.

The enormous oil revenues can have serious repercussions on the international monetary system. The oil-exporting countries should quickly attempt to invest in their own economies. And whatever investments take place, the absorptive capacity of the economies is likely to lag behind the funds available. In the early 1970's the average yearly investment in twelve Arab countries reached $8,450 million. There were variations among countries. Iraq, for example, invested 89 percent of its oil receipts for development financing. The time has come, and the need is pressing, for all the oil countries to undertake investment. The Arab countries' share of world oil exports is about half. In 1972, it was around 700 million tons, as against a world export figure of 1,290 million tons. Exports were distributed as shown in Table 18.

Statistics show that the Arab countries have big oil reserves, which are estimated (as at 1972) as in Table 19. These reserves represent more than one half of the world's reserves.

All this wealth represents a tremendous potential for the Arab countries. But only if investments in productive

Table 18

	In million tons
Saudi Arabia	261
Iraq	70
Abu Dhabi	51
Algeria	48
Kuwait	140
Libya	106
Qatar	23
Total	699

Table 19

	In million tons
Saudi Arabia	18,806
Kuwait	8,940
Algeria	6,135
Libya	3,992
Iraq	3,891
Abu Dhabi	2,828

facilities are undertaken can this potential be translated into actuality. Of course, the fact that the huge oil revenues have an effect on the world economy calls for a dialogue between the oil-producing and the oil-consuming nations to establish a practicable economic and humanitarian relationship.

At present, the oil funds are either put unproductively into Swiss, British, French and American Banks, or are invested in foreign companies. Thus, the huge reserves, constantly increasing, of such countries as Saudi Arabia, Iraq and Libya, and of the Gulf States of Kuwait, Bahrain, Qatar, Abu Dhabi and Dubai, are often operating to the advantage of Western countries rather than to the advantage of developing economies.

A study published in the Economic Issue of *Al Ahram* on 15 December 1973, which reports the resolutions of the Conference of Arab Economic Ministers, recommends the gradual transferring of Arab funds which are currently in foreign banks to Arab organizations for investment in these countries as well as in African countries. In 1973, about $9,545 million was deposited in foreign banks (excluding private reserves and those in other organizations, estimated to total $10,000 million), distributed as follows:

Table 20

	In million US$
Iraq	1,370
Libya	1,840
Algeria	388
Morocco	348
Yemen Democratic Republic	60
Saudi Arabia	3,310
Kuwait	460
Lebanon	705
Tunisia	270
Sudan	39

All these are invested in non-Arab organizations.

With the increase in oil revenue, reserves will climb to much higher levels. It is unlikely that oil prices will go down. Already, the oil-importing countries have taken this into consideration. And since oil is an input, in varying degrees, in the production of both industrial and agricultural goods, prices of these are rising. Already cement prices are 30 percent up and those of glassware 45 percent up. To this must be added transport costs and the expected rise in prices in other commodities not directly affected by oil, which a general equilibrium economic theory tells us about.

From the foregoing it seems safe to assume that the repercussions on the international economic situation of the oil revenues and prices may still alter many variables in the equation.

Economic crises are likely to occur in some of the industries in western countries in the course of the adjustment period. Economic pressures on the developing countries may take place as a result of the decreased aid from the advanced world. This calls for a reconsideration of the plans for development in the developing countries, including the oil-producing countries themselves.

The oil-producing countries must realize the implications of keeping their reserves liquid (by 1980 they may have one-third of world reserves) or of undertaking financial investments in Switzerland or USA or other countries.

The magnitude of the reserves is already so big that one can visualize a situation where they might be transferred from one country to another because of the reluctance of some countries to hold reserves. This is a precarious situation which could eventually lead to a serious situation such as the gradual freezing of reserves by some or all countries.

Failure to invest these vast sums in productive facilities will result in a dissipation of these resources. In an article in *The Guardian* of 1 September 1975, the author begins by saying that when the oil-exporting countries quadrupled the price of oil in 1973–4 the doom-mongers had a field day. He then argues that the oil-exporting countries, by investing and lending their massive surplus abroad, have enabled the oil-importing countries to cover their external deficits without excessive pain. In 1974 OPEC members recycled $56 billions to countries saddled with large oil deficits. In 1975 they were expected to have recycled around $40 billions. OPEC is reducing its current account surplus by spending more on imported goods. In 1974, OPEC invested $38 billions of the $56 billions in just three centres – the US, the UK and Eurocurrency markets. The other industrial countries who were initially unsupported simply entered the US and Eurocurrency markets and borrowed. For developing countries, the situation of secondary recycling was not entirely easy, despite OPEC contributing over $4 billions to the World Bank and the International Monetary Fund to help them lend to the Third World.

In view of all this, perhaps the way out of this impasse is to encourage economic development in the Arab region. This is reasonable, possible and necessary. Investment outlets for the funds are abundant in the developing countries, and the funds would be insulated against the ups and downs of the international monetary system. There are, of course, other important conditions to be worked out, such as ensuring that investment takes place in profitable enterprises, and working out clear-cut agreements between the host countries and the investing countries.

The time has come when a development-oriented approach in the Arab region is a necessity.[1] The countries must realize that the benefits are mutual. An increase in production will result in real gains for all rather than the inflated figures which would characterize financial investment of such magnitudes. The investments referred to here as desirable are not isolated projects such as petrochemicals in Kuwait, or oil pipelines in Egypt, or the Dry Basin Project in the Arab Gulf, but are comprehensive investments based upon scrutinized scientific and technological data, allowing for all the linkages,[2] complementarities and longer time horizons.

Already the Arab countries and the countries of the Third World are facing rapidly rising food import bills. Just as a simple illustration Table 21 gives a few figures for three countries.

Unless an increased proportion of the resources (land, labour, capital and expertise) of the Arab World as a whole is quickly channelled into investment in agricultural and food production, the countries will find themselves in a position where future developments will be more and more costly. Of particular significance in the Arab region is the possibility of investments in fertilizer production, utilizing

1 I shall deal with this subject in my final chapter.
2 *Linkages* refer to the bearing investment could have on other investments.

Table 21

| | Cereal imports (in million US$) | | |
	1969	1972	1974
Algeria	40·1	74·4	288·2
Saudi Arabia	55·3	62·9	158·5
India	417·6	99·9	981·7

Source: *FAO Trade Year Book* Vol. 28, 1974

the natural gas which is at present flared up and wasted. This could have an enormous impact in terms of satisfying the needs of the region as well as providing an exportable surplus. As a result of the rise in food prices, an increasingly bigger proportion of each country's resources, in real terms, is being devoted to the provision of food. The increase in this proportion reduces the availability of resources for investment in productive facilities, in services, education, transport, and other sectors contributing to the gross national product. Furthermore, growth will not be inhibited in the countries of the region alone, but the effects will be transmitted to other countries of the Third World. In the context of the world economy the effects are interdependent, and what happens in one region can have direct and indirect effects on other countries or regions. There are economic relations between the Arab countries and other countries in the Third World. Trade relations are not limited to oil, but the more industrialized Arab countries export manufactures. Furthermore, big imports of food by the Arab countries can have an effect on world prices, affecting in turn other countries.[1]

The countries of the Third World, including the Arab countries, are subject to economic and political pressures. Food is not apolitical: additional pressures on the Arab countries could be exerted.

The resources of the Arab countries are vast. Oil revenues

---

1 The food consumption base of these countries is low. Accordingly, with the rise in incomes and purchasing power, the effective demand for food is rising steeply.

are enormous. It is clear that the Arab oil-exporting countries are anxious to preserve their wealth and that the price of oil is not likely to go down to the old low levels. In the long run, it may go down somewhat as a result of the development of alternative sources of energy. The connection between a rise in the price of oil and a rise in the price of foodstuffs is obvious. Not only is oil an input in agricultural production, affecting, for example, machinery, fertilizers and pesticides, but also a higher price of oil means higher transport costs of food.

The oil revenues, enormous as they are, will continue for a limited time, when the demand for oil will not grow as fast as in the past and when increases in OPEC imports will reduce their surpluses. Some of the oil-exporting countries have the highest *per capita* income in the world (for example: United Arab Emirates and Kuwait). But *per capita* income is not at all, especially in this situation, a satisfactory indicator of development and growth. Only through carefully planned investment will oil lead to wealth in the long run. Idle funds are not only unproductive, but they erode in value. Yes, oil is precious. It is gold, as some say. But paradoxically, gold could lead to starvation. It did to King Midas in the old fable!

Part Three

The International Scene

Chapter
Nine

The World Food
Conference:
The Issues and
the Resolutions

Somehow, rather too suddenly, the world woke up to a Malthusian apocalypse in 1973. It is curious how suddenly the "discovery" was made that there is a serious food problem which is of concern to the world as a whole. In a speech I delivered on 28 March 1974 in Colombo, Sri Lanka, I suggested that the world may have been so preoccupied with economic growth since the Second World War, and with the tremendous strides in technology and science, that food production appeared a commonplace matter, requiring no initiative. In reality the food crisis was not born full-grown. As was shown in the first two chapters of this book, there was a cumulative process: a deceleration in *per capita* food output in the hungry nations of today. The recent disappearance of large North American food stocks and the tripling or quadrupling of the prices of some grain varieties were critical signals.

The developing countries, busy as they may have been with development plans, or talking of foreign exchange gaps, or speaking of the familiar growth theories, found themselves, also suddenly, faced with a new set of problems and a worsening of the old problems: high prices of agricultural inputs, lack of marketing and credit facilities, general inflation, unfavourable terms of trade for the majority and, above all, soaring food prices. Many of them have always in the past subsidized their food prices in order not to reduce the

already low levels of real wages. To subsidize the unprecedented prices of food prevailing in 1973 – this alone entailed
bankruptcy.

For these reasons the Fourth Conference of Heads of State
of Non-Aligned Countries, held in Algiers from 5 to 9 September 1973, called for the convening as a matter of urgency
of a conference on the food problem at ministerial level.
In the same month, Dr Henry Kissinger, the United States
Secretary of State, in his first speech to the UN General
Assembly, made a proposal for a UN World Food Conference under UN auspices. On 17 December 1973, a resolution was adopted by the UN General Assembly to convene
a World Food Conference in November 1974. The gravity
of the world food situation was vividly portrayed by the
United Nations Secretary-General, Kurt Waldheim, when
he said: "The unprecedented growth of the world's population is compounding man's difficulties in feeding himself.
The time at our disposal is very short. The world's food
production has barely kept pace with the population increases. Our goal is not mere survival but a life of dignity
and peace with hope for each new generation to improve
the conditions of life for the billions of men, women and
children who will inhabit the earth in the coming decades."

I delivered about twenty speeches from the time preparations for the Conference began until 16 November 1974,
when the Conference ended. These addresses were given
in different parts of the world and very often the audience
held views different from others.

Since assuming my responsibilities, I have formed my
own conception of what we should achieve. I always tried
to stress the magnitude of the food crisis: that it is no longer
a matter of a famine here and a famine there. During the
first session of the Preparatory Committee for the Conference, which was held in New York in February 1974,
I reminded my audience that malnutrition and starvation
are already there, and that perhaps when the Conference
itself was held, other countries would be added to the list.
The point I emphasized was that this Conference, which

was called at short notice compared with similar conferences, must initiate action.

The second session of the Preparatory Committee was held in June in Geneva. Not only did it discuss the Assessment Paper but at the same time it paved the way for the preparation of the second major document of the Conference: "Proposals for National and International Action". Our main goal was to get the Assessment Paper out of the way as quickly as possible so as to concentrate on action.

In July, I reviewed the situation in my address to the FAO Council held in Rome (15 to 19 July 1974) as well as in my speech to the third Preparatory Committee held in September. I expressed my hope that in the course of the preparatory phase, as well as at the Conference itself, attention would be concentrated on clear-cut fields where progress could be achieved by effective action. I emphasized the need for a world food policy. A country may have an agricultural policy, but how many have a food policy? One cannot be dogmatic about any one food policy. Clearly there cannot be a uniform food policy for all developing countries, development being an integral and organic process. One may therefore not state dogmatically what a particular country should or should not do to solve its food problem. Nor is it permissible, strictly speaking, to separate national policies from international action. Nevertheless, a pattern of consensus has emerged from analysis and discussions carried out which broadly could be called a world food policy. I stated and repeated on several occasions that this policy must rest on three main pillars:

(a) increasing food production in developing countries;
(b) improved consumption and distribution of food;
(c) establishing a better system of world food security.

That increased food production is a desideratum, there can be no doubt. The relevant issue, it was felt, is *how* to produce more. I said that it is a more healthy approach to analyse the obstacles to higher productivity than to talk of the determinants of growth. For instance, how can it be expected that high-yielding seed varieties should achieve their full potential in a situation where fertilizer prices are shooting up, or where there is a scarcity of fertilizer? How can the developing countries produce more if agricultural machinery is not available, or is available only at prohibitive prices? How can we expect increased production of the magnitude we are thinking of unless more land is reclaimed? Recent land development projects in some developing countries have shown that the cost of bringing new land under cultivation can be very high, and beyond the capacity of most developing countries. For instance, in Kenya the cost per acre was $US 173, where the project included dams, reservoirs, canals, flood control, power distribution, and so forth. FAO has estimated that in south-east Asia, South America and tropical Africa, the potentially irrigable area is 260 million hectares; the cost of developing these areas has been estimated at $980 per hectare, or $254,800 million in total.

It was felt that we must be concerned with issues such as those, as a first step towards finding solutions. It is no good to talk of the miracles of the Green Revolution. One is closer to finding a solution if one asks: why has the Green Revolution faltered? It is then that one's attention is drawn to the infrastructure costs needed to make this revolution self-sustaining. High-yield seeds may be fine – for one season – but how can one ensure that there are more available? Investments, and heavy investments at that, are needed in production, education, research and distribution. The farmer will not be able to buy those seeds without credit facilities and the proper institutions that ensure the delivery of seed to 2-acre or 5-acre or 10-acre or n-acre farms. It is only by examining these problems and obstacles that one can come to grips with the brutal fact that, at the end of the decade, of the 140 countries which could benefit from such seeds, only 25 are using them on a significant scale, and of those only 8 have the required infrastructural and other prerequisites.

Those issues were highlighted throughout the Conference

preparations, and I always emphasized that a discussion on productivity in the agricultural sector simply cannot be divorced from rural development. One must first recognize that human resources are the most precious resource the developing countries have. Unless these resources are effectively mobilized, no success can be achieved.

As regards the other components of a world food policy, I indicated in my speech to the third Preparatory Committee that the question is not simply one of food aid, or of stocks. It is not a transient phenomenon. Food aid, better arrangements for emergency situations, price stabilizations and stock arrangements must be viewed on a continuing basis as integral components of a continuing food policy that seeks to balance supply and demand. In a sense this looks like a long-term policy. Well, it must essentially be both longterm and short-term. The long term is emphasized to avoid the sort of situation one expected to find when the world stood helpless until a famine actually started, and the response came too late. But the short term could not be ignored.

Even if and when a long-term strategy is evolved, there will be from time to time acute scarcities in some countries or some regions due to crop failures or climatic factors. A long-term solution must incorporate such short-term measures as would be necessary to cope with emergencies, remembering that emergency assistance and disposal methods as currently practised are inadequate and continue to dry up.

The World Food Conference met from 5 November to 16 November 1974. It marked the first UN-sponsored intergovernmental meeting at ministerial level on world food problems since the Hot Springs Conference in 1943. As such it offered a unique opportunity to world leaders to take bold and decisive action towards elimination of hunger and poverty in the world.

It was expected that there would be differences of opinion in a gathering of representatives from 130 nations as well as many others from non-governmental organizations. The general debate of the Conference was conducted in 11 plenary meetings. The Conference was addressed by the representatives of 104 States, 4 liberation movements, 9 United Nations bodies, 7 international organizations, and 4 non-governmental organizations. The debate reflected the serious concern that the eternal problem of famine and hunger which has haunted men and nations throughout history has now reached an unprecedented scale and urgency, and that it would only be dealt with by concerted world-wide action.

All in all the Conference adopted 22 resolutions, ranging from broad objectives and strategies of food production to specific topics such as fertilizers, pesticides, seed industry development, an improved policy of food aid, an early warning system on food and agriculture, and the establishment of an International Fund for Agricultural Development.

On the subject of increasing food production in developing countries, it was agreed that increased production in the developed countries would have a crucial role during the next few years, before it was possible to build up sufficient momentum in production in the needy developing countries themselves. There was full agreement on the urgent need for a massive acceleration in the increase in food production in the developing countries, which had fallen short of the 4 percent annual average increase called for in the United Nations Second Development Decade Strategy, and had also fallen short of the goals specified in many national plans. Whilst the need for greater self-reliance by the developing countries was stressed, it was agreed that increased foreign assistance was indispensable at the present stage of their development particularly because agricultural modernization depended so heavily on the use of inputs that at present had to be imported mainly from abroad. There was broad support for the estimate in the Conference Document that foreign assistance for agriculture in the developing countries should be raised from the current level of about \$US 1,500 million to about \$US 5,000 million a year by 1980. Several speakers stressed the potential role of the newly increased earnings of the petroleum-exporting countries. In recognition of the need for substantial increase in investment in agriculture for increasing production, a resolution was passed for the establishment of an International Fund for Agricultural Development. Developed countries and those developing countries that are in a position to contribute to this fund were asked to do so voluntarily.

On the other major topic, of improving consumption and nutrition, the Conference stressed that while more food must urgently be produced, the problem of its better distribution and of improving the nutrition of vulnerable and deprived groups, particularly young children and pregnant and nursing mothers, would remain crucial. Consequently, considerable stress was laid on the urgent need for measures to improve the quality of the diet for special nutritional programmes in favour of these groups. A first step was to understand the problem better so as to be able to develop methods for combating it. On the resolution pertaining to nutrition, the Conference recommended that the World Health Organization, the Food and Agriculture Organization, and UNICEF should establish a Global Nutrition Surveillance System, and should arrange for an internationally co-ordinated programme in applied nutritional research.

On the subject of world food security and food aid, it was generally agreed that this forms one of the main pillars of a world food policy. The basic objective of such a system was to ensure that the world has enough food at all times and at reasonable prices. The Conference passed a resolution that a Global Information and Early Warning System on Food and Agriculture be established in FAO. It urged the participation of all countries, especially those large ones that might significantly affect the world situation. On food aid, all were agreed that despite efforts to be made to step up food production in the developing countries, food will have to be transferred for the foreseeable future on a substantial scale from where it is in surplus to where it is in shortage. The Conference recommended that all donor countries accept and implement the concept of forward planning of food aid, make all efforts to provide commodities and/or financial assistance that will ensure in physical terms at least 10 million tons of grain as food aid a year, starting from 1975, and also to provide adequate quantities of other food commodities. On the issue of stocks, there was broad agreement on the need for a co-ordinated system of stock-holding to ensure that the movement of stocks does not lead to shortages.

Although many delegates to the Conference recognized that it was not a trade negotiation forum, there was broad agreement that the subject of trade cannot be excluded in a conference on food. The Conference called on governments to co-operate in promoting a steady and increasing expansion and liberalization of world trade with special reference to food products. It requested all governments to co-operate towards a progressive reduction or abolition of trade barriers and all discriminatory practices, taking into account the principle of most-favoured nation treatment as applicable in GATT.

Finally, in appreciation of the complex nature of the world food problem, which can be solved only by an integrated multi-disciplinary approach, the Conference called upon the General Assembly of the UN to establish a World Food Council, at ministerial or plenipotentiary level, to function as an organ of the United Nations reporting to the General Assembly through the Economic and Social Council. The Council is expected to be a co-ordinating mechanism to provide overall integrated and continuing attention for the successful co-ordination and follow-up of policies concerning food. The World Food Council held its first meeting from 23 to 27 June 1975 in Rome.[1]

Has the World Food Conference averted the catastrophic danger of the food crisis? Or can it? This is the subject to which we address ourselves in the next chapter.

---

1 See the next chapter.

Chapter
Ten

What after Rome?

The Rome Conference attracted the attention of the whole world. True, it was not the first gathering of representatives of nations to discuss the food issue: it was preceded by congresses and meetings to discuss the same problem.[1] But it was the first UN-sponsored intergovernmental meeting at ministerial level on world food problems since the Hot Springs Conference in 1943. Moreover, it was convened at a time when the food problem was essentially a crisis of an unprecedented magnitude. For these reasons, it is no wonder that everyone had high hopes of the Conference and of what would emerge from it. The urgency was the greater because, as we have said, there is no substitute for food, and food cannot be postponed.

It was for this reason that immediately upon assuming my duties as Secretary-General of the Conference I said, and constantly repeated afterwards, that it must have action as its aim.

In my keynote conference speech I stated that the deeper causes of the world food problem lie in rural poverty and in traditional, as opposed to modern, agriculture in the developing countries. History has proved that poverty is neither inevitable nor self-perpetuating. The developing

---

1 For example, the two World Food Congresses held in 1963 and 1970.

countries have emerged into nationhood after prolonged struggles, but they cannot change their ways of life without help from the developed countries. Our objectives were summarized in the three main pillars which I referred to in the previous chapter. These objectives are not entirely new; what was new was the emphasis on an integrated approach. I also emphasized in the speech that without an effective following up to the Conference's resolutions our efforts would be nullified. I reminded my audience that not all countries are advanced and not all countries are oil-producing. The vast majority of the countries of the Third World are importers of both energy and food. Today the poorest nations of the world have a population of one thousand million with annual incomes averaging less than 200 dollars per head. The message I wanted to drive home was that the goals that have been set before this Conference cannot be reached without more effective co-operation between the world's fortunate one-third and its unfortunate two-thirds.

Was the Conference successful? An answer to this question depends upon an examination of three phases: the immediate, the short term, and the long term. Perhaps in the longest of long runs, the problem will be overcome, but because of the nature of the problems one must not ignore the immediate period and the short term.

I delivered my main speech to the Conference on 5 November, but I felt it necessary to make further remarks at the end of the Conference. I said that the first accomplishment of the Conference was the generation of widespread interest in and concern for the problems of hunger and malnutrition. Even the chronic problems of malnutrition to which somehow the world had regrettably begun to reconcile itself had come into sharper focus. Whatever differences of viewpoints there were within groups, everyone recognized that there was an immediate humanitarian problem which must be solved. Next, the Conference had accepted the basic conclusion of the Preparatory Committee, namely that the solution of the food problem requires co-ordinated action on three important fronts: to increase food production, especially in the developing countries; to improve consumption and distribution of food; and to build a system of food security.

Thirdly, regarding the objective of increasing food production, I stated that the resolution to set up an International Fund for Agricultural Development must be regarded as a notable achievement of the Conference. Fourthly, the decision of the Conference on food information and food security represented in my view another landmark. For the first time, the foundations of a food security system had been laid. Fifthly, the recommendation that all donor countries accept and implement the concept of forward planning of food aid, and make all efforts to provide at least 10 million tons of grain as aid every year, was another significant step. This not only insulates programmes of aid from the effects of fluctuations in production and prices, but also provides a more positive policy framework for such programmes in future.

But I indicated that there was one important area where the action of the Conference fell short of my expectations. This concerns the short-term problem. I stated: "In the current situation of food shortage and high prices, the most seriously affected countries need at least 7 to 8 million tons of additional food grain in the next 8 to 9 months. Unless we can provide this grain quickly, a large number of people will face starvation despite all the resolutions and the decisions of the Conference."

I ended my remarks by saying: "Let us all remember that the resolve we have taken that within a decade no child will go to bed hungry, that no family will fear for its next day's bread, and that no human being's future and capacities will be stunted by malnutrition, is a solemn pledge of the entire international community. By this pledge, history will judge the adequacy of our policies and actions." I added: "Let us hope we shall not fail."

So much for my comments on the resolutions of the Conference. But what will happen after Rome? A few months after the Conference and after the prolonged drought in the Indian sub-continent and parts of Africa, the situation has somewhat improved.

What about the short period? If the Conference resolution on a food aid target of 10 million tons a year for the next few years is implemented, then mass starvation can be averted. But will it? Is the task easy? Clearly, it is not easy. The internationalists may argue for a political approach, fitting food into the larger nexus of world inter-dependence. But food cannot be produced without producers. In some of the big producing countries, the producers' standpoint is quite different from that of the internationalists. Their goal is to protect themselves and to try to achieve their own maximum advantage. This does not mean that their policies necessarily contradict those of the internationalists, but they do not always coincide either. And among the contributing nations, not always co-operative with one another, how are the tasks of contribution to be distributed?

The Conference's resolution on food security, if implemented, will solve many of the problems of the past two years. Essentially, the system does not purport to prescribe rigid prices but aims at avoiding extreme fluctuations both in prices and in the quantities supplied. Prices will still fluctuate, and there is nothing wrong with that as long as the fluctuations are not wide and disruptive; after all, prices have fluctuated ever since there were prices. But the precise question is: how easy is it to achieve comparative stability? The answer is that there are many difficulties that beset such a system in practice. One prerequisite for its functioning is the availability of information on harvests and on needs. It is likely that some countries will push hard for broader exchange of information. Given the present situation and the secrecy surrounding output figures of some major countries, this is not easy to achieve. Furthermore, stockbuilding has a depressing effect on prices. There is therefore the question of what level of stocks is acceptable. And acceptable to whom? It may be acceptable to one

97

country but not to another. This is clearly linked to the price range to be agreed upon. A serious problem is the question of storage costs. It is difficult to imagine a situation in which the producers are willing to accept lower prices and also bear storage costs, and it seems that a formula for these must be developed.

A factor which cannot be ignored in all this discussion on food security is the timing of stabilization. It is simply unrealistic to talk of stabilizing prices in a period of rising prices. Who would want to stabilize them? The recent decrease in the prices of cereals may, however, improve the situation.

There may be no choice among policies for many countries, but to admit this is not to argue that there should be no policies. So far no country seems to have developed a major food policy as distinct from an agricultural policy. Even if aid flows, even if a food security system is developed, the food problem is such that it cannot be solved except by increasing production, particularly in countries which are trying to utilize new land or water resources – the poor countries which are at the centre of the food problem. As mentioned in the previous chapter, this was clearly recognized by the Conference.

But how can one achieve this? Let us remember that resolutions are no substitute for resolves. The basic requisite is a concerted international programme to meet the strains that will eventually fall on the international community. The present political state of the world does not seem to conform to such a requisite; the divisions are greater than they used to be. The so-called traditional donors who in the past shouldered the aid bill are now examining their balance sheets. The new donors do not seem to be in a position yet to assess the situation and the priorities.

Given this, can we expect the international community to take the necessary action? May we, indeed, yet speak of an international community prepared to take any action whatever for the benefit of the world as a whole? To what extent have we advanced beyond the stage of individual countries each seeking its own immediate advantage, indifferent to the welfare of its neighbours?

Perhaps it is not our business to attempt, here, any answers to these questions. Perhaps we must leave them to the passage of time. But how much time have we? The nature of the problem is such that waiting may be synonymous with deprivation and starvation. It does seem that it will take some time for countries to realise that the problem cannot be solved by a handful of fertilizers here or a tractor or two there. Things such as the Green Revolution may lead policy-makers to believe that the problem has disappeared or can be deferred. But what one should concentrate on now are the facts about the Green Revolution rather than its glamour in a few countries. We must concentrate on the costs of producing more and what this entails. The world now utilizes 80 million metric tons of fertilizer each year. By 2000 the figure must rise to 250 million metric tons. To reach this level, new plant capacity alone will entail expenditure of 8 to 9 billion dollars. Even if this is achieved, I firmly believe that, because of the characteristic lumpiness of investment in fertilizer plant, sharp fluctuations of supply, and consequently of prices, may occur. Accordingly, there is a continuing need for a body at the international level to channel, multilaterally, supplies of fertilizers to developing countries in periods of relatively short supply and/or high prices.

The payment for basic foodstuffs is the start of the problem. The bigger question is whether the world can actually produce enough to meet the total demand. The inflation which is besetting the world economy means that one has to constantly revise the cost bills upwards – and this means that the food problem will become more and more pronounced.

The UN World Food Council, the highest policy-making body on food issues, which was entrusted with following up the World Food Conference resolutions, did not demonstrate when it met in June 1975 that the international community was unified in its approach to tackling the food problem. In November 1975, I was invited to speak to the Food and Agricultural Organization's Eighteenth Conference, one year after the World Food Conference had ended. I stated that whilst there were some encouraging signs that the international community is beginning to channel more resources into food production, the world was still very far from developing, or evolving towards, a secure and flexible world food security system in which, through co-ordinated food policies, through improved arrangements for emergencies and food aid, and through more harmonious trade and production policies, the world could achieve a better balance between demand and supply of food. Discussions on this subject in existing fora have not progressed. Food aid in 1975 reached 9 million tons of cereal grains. This is below the target of a minimum 10 million tons which the World Food Conference had set. I reminded my audience that, furthermore, we are not solely concerned with a figure of 10 million tons: we have set ourselves a new policy framework in which food assistance will shift from a fluctuating surplus disposal programme to more important objectives of nutrition, employment, and development, in the context of a long-term framework ensuring a reasonable degree of continuity in physical supplies.

At its second session, held in June 1976, the World Food Council had a comprehensive agenda, dealing mainly with food production, food aid, and food security. That session went a step further than the first had done. In general, it was clear that although the study of many technical problems had been advanced in various discussions, there had been little real progress in evolving a dependable system of food reserves and food security; for this, the secretariat presented proposals, and it was agreed that some technical details of the provision of reserves needed further elaboration. Steps towards a system will be a major item on the agenda of the Council's third session.

An improved policy for food aid was generally recognized to be needed. The concept of forward planning of food aid was supported by most delegates present at the meeting, and was considered necessary for achieving the objectives of nutrition in the long term.

On the subject of increasing food production in developing countries, the Council had before it a secretariat paper considering those countries to which priority should be given; this proposed certain criteria and guidelines, but it was felt that these needed further work. This topic, too, will be high on the agenda of the third session, where clearer directions and sharper priorities should be defined so that a practical and realistic strategy may be developed for an increase in food production. It was observed that although external assistance to agriculture had increased since 1972, and a higher priority had been given to increasing production in both developed and developing countries, the factors underlying the need for the increase had not changed.

The third meeting of the Council, which took place in the Philippines in June 1977, discussed nutrition and food trade. An effective programme to reduce malnutrition in the world has yet to be undertaken, and international trade in food has been much discussed without any real progress towards liberalization.

It may be that one task of the World Food Council will be to inaugurate, on the problems of agriculture if on nothing else, a unified international community which alone can solve these problems. If such a community does not yet exist, the Council must nevertheless hold it constantly as its aim, undiscouraged by difficulties; it must never allow itself to become a mere forum of debate, but must insist on action, both immediately and for the future. Temporarily, because of present urgent necessity, we may have to modify our strategy. We may have to seek a situation in which donors will be encouraged to give aid by the prospects of resultant benefit to themselves. If the oil-producing countries, most of which are net importers of food, start to work out a big plan at a regional level whereby food production is encouraged in neighbouring countries by joint ventures, then this might be a way out of the impasse. The setting up of fertilizer plants would remove one of the most important bottlenecks: shortage of fertilizers. The realization of the food problem and its ramifications such as inflation, monetary instability and general speculative activity may be an incentive to this end. It is a reminder that the food problem must be elevated above the sectional problem of agriculture. It is admittedly part of the general problem of development, and its long-term solution is intimately tied to the achievement of general development and progress. But because it is so urgent and so fundamental to all other efforts, it must be treated as a standing emergency for the coming years.

If the delegates to the Rome Conference appreciate all this and do something, then something will be achieved. The Romans will not need to pray again to the gods for food!

Chapter Eleven

An Arab "Marshall" Plan

We are only half way through the decade of the 1970's but things on the economic scene are moving fast, so fast, indeed, that it looks as if our familiar economic theories need rethinking.

There is much current talk of the enormous capital resources of the oil-exporting countries. We read practically every day about the difficulties attendant upon the international monetary system. Indeed, the financial side of economics has captured the attention of every writer. Some argue that the accelerating accumulation of Arab funds is disrupting the whole system.

An economic crisis there may well be. There is inflation of an unprecedented degree at an unprecedented rate. There is a food problem; or rather a food crisis. There is a rise in the prices of manufactured goods. It appears that the speed with which events are moving outstrips the devising of remedies. All this is reflected not only in a frenzy of articles and speeches but more concretely in the number of important world conferences, on a big scale. In a very short span of time, the international community gathered for an Environmental Conference, a Population Conference, a Food Conference, a UN Special Session on Raw Materials, and a Law of the Seas Conference. This is unusual, if past experience is any criterion.

But events themselves seem unusual. What are the facts? That there is a massive flow of new resources is unquestionable. In situations where events assume such colossal magnitude it is well to guard against misconceived allegations. Some countries seem to have attributed all their economic evils to one cause.

A detached view of the issues involved might help us to escape from this impasse. There is a problem, and a big problem at that, which afflicts the industrialized countries in one way or the other. Likewise, there is a problem of which the developing countries might rightly complain. In between there is a series of problems or conflicts. It would be misleading to talk of inflation as the cause because the facts tell us that there are marked differentials in the rates of change in prices. Our old dichotomy between prices of primary commodities and prices of manufactured goods does not help much. Just look at primary commodities. These include food, agricultural raw materials, and minerals. In the past two or three years strange things have been happening. Whilst world prices of some food items quadrupled, prices of some raw materials or minerals remained stationary or decreased. Even if one were to take one sub-category such as minerals, a similar degree of discrepancy could be traced.

The one undeniable fact is the increase in liquidity of the oil-exporting countries. It is a stark fact that an increasing liquidity or money supply without a compatible increase in the availability of goods and services results in a general rise in prices. How can this be overcome? Should these funds find their way to the industrialized countries, where the shares of giant companies are bought partially or wholly? Is this a proper channel for utilizing the Arab funds? What if those foreign countries subsequently nationalize these companies? Surely, it seems that the proper solution is investment in the developing countries of Africa, the Middle East, and Asia. Investment in a developing country may have a long-term characteristic. There is nothing wrong with that, because the solution cannot ignore the long term.

This is the long-term aspect of the solution which does have its bearings on short-term policies.

Economic history might give an insight. Could it be that we need such a thing as the Marshall Plan of the late forties? If the Marshall Plan is any guide, then it pays to examine what in essence this plan aimed to do and in what circumstances it was carried out. The Marshall Plan of 1947, called after the then US Secretary of State George C. Marshall, aimed at helping European economies to recover after the devastation of the Second World War. When the war was over, a scarcity of resources remained. Only the United States had not been physically devastated by the war. In Europe there was need to rebuild, to develop, and to make up for scarcities. The expression that floated in the papers those days so widely was the "dollar shortage" or "dollar gap". Technically this meant that the demand for dollars from outside America, to finance needed imports, was much greater than the supply of dollars. The Marshall Plan, with far-sightedness and sound economic reasoning, stated that aid to Europe should not be "on a piecemeal basis as various crises develop" but "should provide a cure rather than a mere palliative." The plan aimed at development of those countries by supporting needed imports of equipment and supplies, and by encouraging internal measures to promote financial stability. Each country was to formulate a plan for the four-year period 1948 to 1951, proposing measures to increase production and reduce balance-of-payments deficits. The plans were to be reviewed and co-ordinated by the OEEC and screened by the Economic Co-operation Administration, to ensure that the United States would provide only necessary assistance and to give it a measure of control over the monetary and fiscal policies of the participating countries. Over the four-year period of operation of the European Recovery Programme, the US extended $11·4 billion of aid to Europe, almost 90 percent of which was in the form of outright grants. The result of the plan was encouraging. A substantial expansion of European production was achieved. In 1951 industrial production in

105

Western Europe was about 40 percent greater than in 1938, that being a record year.

A study of the Marshall Plan shows that the USA did not aim solely at benefiting the European countries, but also at benefiting itself. USA exports in 1947 totalled $20 billion whilst its imports were valued at $8·5 billion. Clearly, this state of affairs could not continue in the long run. The USA, by giving aid to friendly countries, contributed to the achievement of equilibrium, the benefits of which accrued to all.

There seems to be some resemblance between the period of the dollar shortage of the late forties and what may now be called "petro-money shortage". In the late forties it was America which, relative to the other countries, had greater liquidity. Europe had a shortage. Today oil-exporting countries are generating funds or liquidity, and other countries, especially the developing countries of Africa and Asia, are taking the place of Europe of the late forties.

It could be similarly argued that the setting up of a fund for the developing countries will contribute positively not only to recipient countries but also to the contributors. The developing countries are, and will remain for a long time, the main producers of raw materials such as oil, iron ore, copper, manganese, lead, rubber, oilseeds, and so forth, all of which are of strategic value in international development.

The basic idea is simple: the developing countries possess vast land and human resources; they need capital, technology, and appropriate managerial skills.

The idea of an Arab "Marshall" Plan had been revolving in my mind before I became Secretary-General of the World Food Conference in February 1974. I thought that one might start with an important but less ambitious project. The preparation for the Conference offered me an opportunity to launch the idea of an International Fund for Agricultural Development (IFAD). So convinced was I of the desirability, if not the necessity, of having such a fund that I have been approaching at the highest level certain OPEC countries with a view to a fund of US $1,000 million. I always felt that such a fund would constitute a major addition to the resources now going into investment in agriculture. Since a large number of the present loans to agriculture are not on soft terms, concessional loans from the Fund could considerably increase its effect. Moreover, a specialized international fund solely for agricultural development could draw the attention of all financial institutions to the importance of agriculture; to that extent, the Fund might indirectly affect lending-policies.

The Secretariat of the World Food Council has been busy preparing several meetings of interested countries and working groups on the establishment of the International Fund for Agricultural Development (IFAD). A working group has held several sessions to decide the operational and technical details of the Fund, and has made good progress. The contributors to the Fund are expected to be the OPEC countries and the advanced countries. The concept of its governing board is that the OPEC group, the advanced countries, and the recipient countries should each have one third of the total votes. Thus, for the first time, a fund will exist in which two-thirds of the voting power will lie with the developing countries and two-thirds with the donors; this apparent anomaly reflecting simply the fact that the OPEC countries belong to both categories: donors and developing economies. Politically, this seems to be a desirable structure, wherein countries of all economic types can work together for the common good.

Towards the end of 1975 Iran proposed a levy of 10 US cents per barrel of oil, to be used to help the developing countries. In January 1976 this proposal fructified in the formation of an OPEC Special Fund for this purpose, with a capital of US $800 million.

Since the first edition of this book appeared, the International Fund for Agricultural Development has become a reality. In December 1976 the agreement establishing the Fund was opened for signature, and contributions surpassed the immediate aim of US $1,000 million, reaching 1,012 million. A preparatory commission has since been set up to work out details, especially those of policies and criteria for loans.

Not only is the idea itself of the Fund an innovation; so is its approach to agricultural development. The emphases on food production, on reducing rural poverty, and on improving nutrition are examples of a new and realistic approach to rural development. This is wholly in line with the principle that agricultural development must not be considered mechanically; rather, it is rural development that is the objective, with emphasis on social aims as well as economic aims. The translation of these aims into real criteria for loans is no easy task; the complex issues involved will require fresh thinking.

How does all this fit into my proposed "Marshall" Plan? I think it could fit very well. The Plan in my mind could embrace the IFAD, OPEC's Special Fund, and other potential funds for industrial development and for rural services. Eventually, advanced countries may wish to join OPEC. The structure may be sketched as follows:

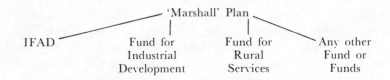

Basic features of the plan

The setting up of a development plan calls for a detailed study covering membership, mode of operations, terms of financing, and so on. These features are technical in nature and it is best to leave it to financial and legal experts to spell them out. Here, only the broad features are outlined.

1 Objective

The objective is to give aid to the developing countries for economic and social development. This can be in the form of either grants or loans, depending on the circumstances. In addition, it is expected to provide technical expertise. In doing all this, the funds may

(a) contribute to the identification of new investment outlets in the developing countries;

(b) direct capital inflow into projects which benefit the member countries;

(c) prepare comprehensive studies of projects;

(d) offer consultative services, and see to the availability of experts and their exchange among countries;

(e) undertake an analysis of technical aid given to international organizations after proper review from the technical organs of the banks (referred to subsequently) through which financing will take place;

(f) undertake the necessary steps to develop technical and managerial skills.

2 The financial and managerial structure

The member countries should decide on the structure, which while being independent could have links with the UN.

3 Resources

Contributions may be made on criteria to be agreed upon, such as GNP, proceeds of exports or others. Initially, the capital may be $3 billion, which may gradually increase to reach $20 billion in ten years. In any case, the magnitude of the development plans of the developing countries will condition the size of the funds' capital.

4 Loans and guarantees

It seems that a workable solution would be for the fund to give loans to the developing countries through a consortium of banks which in turn would be responsible for settlement of accounts. The consortium might embrace a group of fourteen international banks selected from regions representing the developing and the developed countries, and the World Bank or an organization affiliated to it.

These banks would be expected to undertake tasks over and above ordinary banking functions, such as consultative work and project planning. This is important because the developing countries need projects planned and executed on sound scientific principles. For this reason, it is necessary to have a Consultative Committee at the highest level, in addition to whatever services are available from the UN and other organizations.

The terms of financing should be compatible with the economic condition of the recipient countries. Two types of loans may be conceived: soft loans and commercial loans. In addition to this, the funds may give outright grants upon the recommendation of the Executive Board.

One matter which cannot be ignored is the working out of suitable guarantees for loans.

* * * *

The oil-exporting countries have established many funds with the objective of development, such as the Kuwait Fund and the Saudi Fund and others I referred to earlier. In so doing, these countries felt that they had fulfilled their role in promoting development. There are, however, many factors which are lacking, such as technical and planning expertise.

The developed world unjustifiably looks at the oil-exporting countries as responsible for the inflationary situation in the developed countries. The non-oil-producing developing countries look hopefully to the oil-exporting countries, and at the same time they are facing higher prices, as the Prime Minister of Sri Lanka mentioned, expressing the view of all those developing countries which need to import oil.

The Dakar Conference (3 to 8 February 1975), although it was held for the Non-Aligned countries, was effectively extended to embrace the whole of the Third World. The People's Republic of Korea, Brazil, and the Philippines joined. There were eleven developing countries in addition to the fifty-seven Non-Aligned countries, as well as representatives from international and regional organizations, liberation movements, and three developed countries: Austria, Sweden, and Finland. The basic issues were: the evaluation of trade in primary products; the establishment of a new international economic order; co-operation among the Non-Aligned countries in the trade of raw materials; and preparation for the Conference of Developing Countries on raw materials.

There were certain conspicuous features of that Conference.

One view aimed at embarrassing the oil-exporting countries by holding them responsible for the problems of underdevelopment. A proposal was made for the establishment of a fund by the oil countries with a capital of $6 billion, the objective of which should be to stabilize prices of primary products. Some countries were sympathetic with this proposal. Some oil-exporting countries did not actively participate in discussion; others reserved their positions on any resolutions that the Conference was expected to adopt. Another standpoint emphasized the unity of the Third World, including the oil-exporting countries, vis-à-vis the advanced industrial nations. Others called for reconciliations with the advanced countries and the avoidance of any confrontation.

The majority of members, however, emphasized the unity of the Third World and the necessity for establishing a new international economic order, guaranteeing equity to the developing countries.

Two major points seem evident: that the non-oil-producing developing countries feel the effects on their economies of the inadequacy of the requisites for development, and that the economic tension between the developed and the developing countries is increasingly apparent in international gatherings.

The main point I would like to make, taking all this into consideration, is that economic development of the poor countries is now an imperative. At the same time, there needs to be a clearly defined role for the oil-producing countries and another for the advanced developed countries. I believe that the time is now ripe for the Arab countries to state categorically that they are ready to play a positive role in eliminating poverty, and to invite the advanced countries to join them.

I go back to the point I made earlier, namely the necessity of suitable guarantees for investments by the oil countries to ensure a steady flow of funds to be channelled in an organized way according to the documented needs of the recipient countries. Thus, one would overcome the inevitable and endless disagreements and quarrels on who gets what.

The recent world conferences (for example, the World Population Conference and the World Food Conference) have clearly shown the necessity for development in the poor countries and for mobilization of resources for a speedy solution. There are obstacles on the way, but if we mistake not, it seems that the world is becoming more and more conscious of interdependence and is realizing the dangers of economic isolationism. The last three years have been the scene of more than one crisis, including the food crisis. Such crises, fraught with dangers as they are, also bear promise if the international community faces them with an open mind and with a spirit of co-operation.

This is a challenge. But there are reasons for optimism.

Chapter Twelve

A New International Economic Order

If the third quarter of the twentieth century has been on the whole the scene of economic growth, the fourth quarter, already begun, is fraught with difficulties, or, perhaps more correctly, with crises. The proliferation of international conferences in recent years is expressive of an awareness of the growing difficulties, if not the breakdown, of the world economic order.

This order, born after the Second World War, has now been eroded, if indeed it has not totally collapsed. Its systems may have served well, but many of them are now showing their quarter century of age. Neither new economic theories nor empirical studies can replace thoughtful and practical attention to the politico-economic issues that have been developing so quickly.

The result has been the growing economic difficulties faced by virtually all nations: slow growth, recession, inflation, a threat to the international monetary system, and even depression. The gap between rich and poor countries is widening, and the prospect over two or more decades is that four out of every five individuals on this globe will live in poor countries. Numbers alone require that they be not left out of account.

What are the implications of this? The one clear-cut fact is that we live in a world of increasing interdependence. It is therefore necessary to remove the definition of the problems from the usual narrow concepts of one group of countries versus another group of countries. A recognition of the interdependence of our lives makes it incumbent on us to define the problems anew. I suggest that we refer to them as common problems. I would also add that it would be misleading to talk of only one component of the problem, be it military, political, economic, or social. All are interwoven, and it is their unity that should dominate our thinking if co-operation is to reign. But that is not enough. Our agenda for any one year should include items of short-term urgency, and should at least consider short-term policies which contain the potential for evolution into longer-term policies for interchange of resources to enhance co-operative efforts to accelerate development.

In May 1974, the UN General Assembly adopted the declaration on the establishment of a New International Economic Order designed to eliminate the widening gap between the developed and the developing countries. It required developing countries to participate actively, fully, and equally, "in the formation and application of all decisions that concern the international community". And we read further that "international co-operation for development is the shared goal and the common duty of all countries."

I wish to pause here for a moment. It is no secret that the call for a new economic order has startled some quarters in the world. But why should it? Perhaps it is the word *new* that caused a cold reception. To deter any possible spectres, let us come to grips with the fundamental problem. Then the substance will be unveiled, and the spectres be seen for what they are. The fundamental problem is, as I said at the outset, that the international system is breaking down or has broken down, and the suddenness of our grasp of this fact is for the moment irrelevant. It follows logically that a new system should prevail, or let me put it simply: we need a system. It must be created soon, and all countries should participate in its creation. That by itself does not guarantee that all will be happy and content. But perhaps they never will be. What is primarily important is that there should be a consciousness of participation and that there should be visible means for answering everyone's anxieties.

A new system is necessary if the world is to develop significant co-operation and coexistence. For without it there can be only a state of affairs where groups will join other groups, and international problems will be expressed in endless dissensions and tests of strength, resulting in political, economic, and social unrest – perhaps in military conflict.

The relations between the industrialized nations and the Third World have become crucial to international economics. This is a wide subject and I cannot do it justice here. I shall concentrate on the two topics of economic co-operation among developing countries and the interchange of technology, and I shall try to discuss them in the broad perspective I have so far been talking about.

Economic co-operation among developing countries has been receiving increased attention in recent years. Hitherto it has been mainly conceived as a means for elaborating similar or identical positions at various international meetings, especially within the UN, with the aim of various benefits to countries of the Third World.

More recently the emphasis has been increasingly put on active economic co-operation among these countries themselves, in such fields as preferential trade agreements, exchange of surplus financial resources, joint enterprises, and the development and exchange of appropriate technology.

It is my firm conviction that economic co-operation among developing countries will be a major force propelling us towards a new order. But from where do we start? Clearly, co-operation does not mean that any one country should pursue a policy of autarky, but it implies a full mobilization of a country's economic resources.

I envisage some sort of agreement for economic co-operation among Third World countries to which all may subscribe regardless of differences in their socio-economic structures. Differences in the degrees of development among these countries are no barrier. Each should have an obligation to join in: the process of development should not be looked upon as lying in the hands of any single group of nations. At the same time, co-operation among Third World countries should not be a means for producing among these countries any relations of dominance or subordination.

Such co-operation among developing countries will help the process of development; at the same time it will give a sharper delineation to their relations with the rest of the world. This will in turn foster a sense of obligation on the part of the developed countries. Many of the issues we now talk of will become fundamental concerns to developed countries. The developing countries will be in a stronger position in world-wide capital markets. The developed countries should provide for the developing countries stable and efficient markets. Special measures are needed to help developing countries find new markets for the goods and services which they will be increasingly able to provide. And the international community should help the developing countries to exchange, develop, adapt, and manage technology appropriate to their needs. We should not judge from sporadic and limited past experience of inappropriate exchange of technology: for example, technology dependent on relatively scarce factors of production or biased towards the production of luxury consumer goods.

Technology itself is a process. It is not a piece of metal that is being passed from a supplier to a recipient. The stark fact is that it is one very scarce resource in developing countries. If we are going to work in an international environment in which any economy is ever going to prosper, then no country can afford to forgo the concern for development. Development is an expression of the entire history of a developed country: it is the aspiration of a developing country.

The success of technological interchange depends, among other things, on a recognition that it must be in the form of a constant flow. It would be immature to judge from sporadic incidents. What is needed is a recognition that development is a continuous process and is dynamic in nature.

If such an attitude is taken, then a fresh look at technological interchange will show that technology itself is no magic but an important resource which needs to be flexibly and creatively applied, given the dynamic nature of the development process.

At the UN World Conference I emphasized from the start an integrated approach to the food problem alone. When we come to a subject such as technological interchange, an' integrated approach is all the more necessary. The UN Conference of 1963 on the Application of Science and Technology to Developing Countries did not live up to expectations. There will be another in 1979. I now feel that I should somehow sound a note of warning. Unless an integrated approach is envisaged in the full realization that a unified, interdisciplinary, and dynamic outlook is essential, the outcome may leave much to be desired. It must be recognized that development is a process in which 1963 or 1979 do not mean much, but what matters is dynamic and constructive planning for decades to come. No year is a final year where things come to a halt: I recall that when we were making projections to 1985 for the food deficit, a misconception arose implying that 1985 was a stopping place. For this reason our projections had to be given a dynamic element. A year in the future may be the last in a statistical computation, but it may not be assumed to be a stopping place.

This dynamic nature of things is also at the heart of economic co-operation among developing countries, from which the question of technological interchange cannot be divorced. For economic co-operation itself may result in fuller utilization of resources and more economic efficiency in individual nations, and may thus act as a propellant to the process of development. In this case, technology will become less complex, and a sharper focus will be seen.

For this reason, I should like to go back a little to the subject of co-operation among developing countries. I have so far talked about co-operation among all developing countries, but we should not ignore individual attempts at integration. There have been several efforts to secure integration among developing countries: for example, the LASTA, the Central American Common Market, the East African Common Market, the Arab Common Market. Some of these groupings are older than others. But in no case have the practical results been more than modest so far.

The balancing of gains and losses in a regional economic grouping is well known. I do not wish to go into this here, but I would nevertheless like to concentrate on a few aspects relevant to my thesis. In most regional groupings, intra-regional trade represents a small proportion of total trade; in some cases a declining proportion. True, over the long run, cost and trade patterns may change, so that the scope for net additional gains will become greater within a union. But to argue this way is somewhat begging the question. The precise issue is that a union among developing countries should itself be viewed as a vehicle for dynamic change: change in the cost patterns themselves. The problem is usually one where the immediate gains are small and the longer prospects are favourable. But it is these longer prospects that should influence current decisions. Besides, another common difficulty is the realization that member countries in a union or a grouping are unlikely to benefit equally. This question of equitable distribution of costs and benefits is perhaps the most difficult and contentious. Evidence seems to suggest that unless a union is strong enough to adopt fiscal and other measures to distribute gains more evenly, its viability is threatened. Moreover, many attempts at integration among developing countries seem to have been beset by lack of multilateral payment mechanisms. These factors acted as a brake on trade exchanges.

An observable tendency is that in many cases where trade within a regional grouping expanded, trade between that grouping and other developing countries, or trade between groupings, did not seem to expand; in many cases it contracted. This is an important point, since the goal is full co-operation among developing countries. In this respect, it is worth recalling that the relevant resolutions of the UN General Assembly in recent years indicated, among other things, the steps required for the expansion of co-operation at the sub-regional, regional, and inter-regional levels. More recently, the conferences of the Group of 77 on Economic Co-operation among Developing Countries, held in Mexico City from 13 to 22 September 1976, took several decisions pertaining to this subject.

There is one area which, I believe, is of the utmost importance. This is agricultural activity. I hope my remarks here will be considered for what they are worth, ignoring my personal bias and my experience in the field of agriculture and food. The point I want to make is that agriculture in general and food production in particular must be included in regional integration schemes. It is regrettable that this subject has not received enough attention. There seem to be two reasons. Historically, groupings concentrated on manufacturing industries; indeed, the theory of customs unions, when first developed in the early fifties, referred to advanced countries, with manufacturing activity dominant. Next, there is an inadequate understanding of the process of integrating agricultural sectors, and there are difficulties in modifying agricultural sectors by national action planned within a regional frame. The case for the inclusion of agriculture rests on economic, social, and political grounds. It is usually a large sector in a developing country, and therefore exclusion would inevitably lead, according to simple economic logic, to inter-sectoral distortions affecting overall economic development. Specifically, in regard to food, many developing countries are faced simultaneously with increased food imports and under-utilized agricultural resources. Regional integration and general economic co-operation among developing countries can contribute to raising productivity and enlarging the market. This is the more urgent when it is recalled that the dependence of developing countries on a few suppliers of food in the world market is increasing alarmingly.

A word of warning is however necessary in discussion of the integration of agricultural sectors. If integration of these sectors is sought by trade approaches alone, benefits are likely to accrue to those sectors which are already fairly well developed, to the exclusion and damage of low-productivity components of national agriculture. Integration in the context of economic development should not be allowed to result in such lop-sided effects, which are not acceptable socially or politically. There must exist an acceptable and identifiable longer-term perspective for regional agricultural development, in the full realization that this fits in well with overall economic development objectives which must remain the ultimate objective of regional integration and economic co-operation.

It is clear that the core of everything I have said so far is a recognition and a quest for an understanding of the true meaning of INTERDEPENDENCE, which colours our world of today and will characterize our world of tomorrow. The moral issues of colonialism may have occupied minds in the first half of this century. The gap between rich and poor was also looked upon as a moral problem. And there have been a host of other moral problems. But in a world that is becoming more and more dependent on international co-operation – that is, becoming increasingly interdependent – these moral problems and others have practical implications for the well-being of the rich as well as of the poor. The world systems will not forever work for the rich unless they work also for the poor.

Thus, international co-operation for development becomes the shared goal and the common duty of all countries. No country, however rich and strong, can by itself solve the new global problems which the world is facing. The understanding of interdependence poses a challenge. It is not easy. But whatever the word means, it implies that one country's policies, even those traditionally regarded as domestic, are dependent on the actions and decisions of other states.

A thorough grasp that in the final analysis we inhabit ONE WORLD represents a victory for common sense.

Appendix One

Recorded famines, to 1961

Date	Area	Comment
c. 3500 B.C.	Egypt	Earliest written famine reference.
436 B.C.	Rome	Thousands of starving people threw themselves into the Tiber.
A.D. 310	Britain	40,000 deaths.
917–18	India Kashmir	Great mortality. Water in Jhelum River covered by bodies. "The land became densely covered with bones in all directions, until it was like one great burial-ground, causing terror to all beings."
1064–72	Egypt	Failure of Nile flood for seven years. Cannibalism.
1069	England	Norman invasion. Cannibalism.
1235	England	20,000 deaths in London; people ate bark of trees, grass.
1315–17	Central and Western Europe	Caused by excessive rain, spring and summer of 1315. Deaths from starvation and disease may have been 10 percent over wide area.
1333–7	China	Great famine; reported 4,000,000 dead in one region only; perhaps source of Europe's Black Death.
1347–8	Italy	Famine, followed by plague (Black Death).
1557	Russia	Widespread, but especially upper Volga. "Very severe: a great many starved in cities, villages and along the roads." Caused by rains and severe cold.
1594–8	India	Great mortality, cannibalism, and bodies not disposed of. Plague.
1600	Russia	500,000 dead. Also plague.

Date	Area	Comment
1630	India, Deccan	During the time of Shah Jahan, builder of the Taj Mahal, who undertook relief efforts to assist. 30,000 reported to have died in one city, Surat. Drought followed by floods.
1650–2	Russia	Excessive rain and floods. "People ate sawdust." Many died despite Tsar's permitting free grain imports. High grain prices prevented purchase of seed.
1677	India, Hyderabad	Great mortality. Caused by excessive rain. "All persons were destroyed by famine excepting two or three in each village."
1693	France	Awful famine – described by Voltaire.
1769	France	Five percent of population said to have died.
1769–70	India, Bengal	Caused by drought. Estimates of deaths range from 3,000,000 (a tenth of population) to 10,000,000 (a third of population).
1770	Eastern Europe	Famine and pestilence caused 168,000 deaths in Bohemia and 20,000 in Russia and Poland.
1775	Cape Verde Islands	Great famine – 16,000 people died.
1790–2	India, Bombay, Hyderabad, Orissa, Madras, Gujurat	The Doji Bara or skull famine, so-called because the dead were too numerous to be buried. Cannibalism.
1803–4	Western India	Caused by drought, locusts, war, and migration of starving people. Thousands died.
1837–8	Northwest India	Drought. 800,000 died.
1846–51	Ireland	Great potato famines. A million died from starvation and disease; even more emigrated.
1866	India, Bengal and Orissa	Poor distribution of rainfall. 1,500,000 deaths.

Date	Area	Comment
1868–70	India, Rajputana, Northwest, and Central Provinces, Punjab, Bombay	Drought. Famine followed by fever. Deaths estimated at a fourth to a third of total population of Rajputana. In one district 90 percent of cattle died. Shortage of water for cooking and drinking.
1874–5	Asia Minor	150,000 deaths.
1876–8	India	Drought. Over 36,000,000 affected; deaths estimated at 5,000,000.
1876–9	North China	Drought for 3 years. Children sold. Cannibalism. Estimated deaths – 9,000,000–13,000,000.
1892–4	China	Drought. Deaths estimated at 1,000,000.
1896–7	India	Drought. Widespread disease. Estimates of death range up to 5,000,000. Relief efforts successful in several areas.
1899–1900	India	Drought. Extensive relief efforts, but 1,250,000 starved. Another estimate, including effects of disease, 3,250,000.
1920–1	North China	Drought. Estimated 20,000,000 affected; 500,000 deaths.
1921–2	USSR, especially Ukraine and Volga region	Drought, US assistance requested by Maksim Gorki. Despite relief efforts 20,000,000–24,000,000 affected, estimates of death 1,250,000–5,000,000.
1928–9	China, Shensi, Honan, and Kansu	Comparable in extent and severity to great famine of 1877–8, though because of railroads deaths were probably fewer. In Shensi alone an estimated 3,000,000 died.
1932–4	USSR	Caused by collectivization, forced procurements, destruction of livestock by peasants. Estimated 5,000,000 died.
1941–3	Greece	War. Losses because of increased mortality and reduced births estimated at 450,000.

Date	Area	Comment
1941–2	Warsaw	War. Starvation, directly or indirectly, estimated to have taken 43,000 lives.
1943	Ruanda-Urundi	35,000 to 50,000 deaths.
1943–4	India, Bengal	Drought. Burmese rice cut off by war. 1,500,000 died.
1947	USSR	Reported by Khrushchev in 1963. Referring to Stalin and Molotov: "Their method was like this: they sold grain abroad, while in some regions people were swollen with hunger and even dying for lack of bread." (*Pravda*, 10 December 1963.)
1960–1	**Congo Republic (Kasai)**	Caused by civil war.

Source: *Encyclopaedia Britannica*

Appendix
Two

Population of the Arab World
(in millions)

Country	
Algeria	16·1
Egypt	38·0
Libya	2·1
Mauritania	1·3
Morocco	18·0
Somalia	3·1
Sudan	17·9
Tunisia	5·8
Bahrain	0·2
Iraq	11·2
Jordan	2·6
Kuwait	1·0
Lebanon	3·1
Oman	0·7
Qatar	0·1
Saudi Arabia	8·7
Syria	7·1
United Arab Emirates	0·1
Yemen Arab Republic	6·4
Yemen Democratic Republic	1·4
Total	144·9

Appendix Three

Appendices 3 to 6 show the yield of cereal crops in the countries of the Arab World.

WHEAT
Yield in kilograms per hectare

Country	1961–5 Average	1972	1973	1974
Algeria	637	837	512	425
Egypt	2,621	3,091	3,503	3,447
Iraq	701	1,544	828	820
Jordan	671	945	445	1,023
Lebanon	939	1,261	1,100	1,200
Libya	245	379	452	609
Morocco	847	1,081	772	967
Saudi Arabia	1,300	1,200	1,200	1,346
Sudan	1,308	1,248	1,459	1,348
Syria	783	1,335	402	1,061
Tunisia	494	702	776	810
Yemen Arab Republic	1,000	1,080	1,000	1,183
Yemen People's Democratic Republic	2,025	1,071	1,071	1,000

Source: Compiled and Computed from the *FAO Production Yearbook, 1974*

Appendix Four

BARLEY
Yield in kilograms per hectare

Country	1961–5 Average	1972	1973	1974
Algeria	587	921	696	750
Egypt	2,614	2,675	2,685	3,061
Iraq	925	1,452	995	1,027
Jordan	689	561	308	583
Lebanon	968	966	888	1,000
Libya	248	710	714	333
Morocco	808	1,261	616	1,196
Saudi Arabia	1,046	1,600	1,500	1,571
Sudan	–	–	–	–
Syria	877	1,197	112	940
Tunisia	286	613	525	882
Yemen Arab Republic	1,000	1,245	1,034	1,353
Yemen People's Democratic Republic	2,255	2,917	2,917	2,917

Source: Compiled and Computed from the *FAO Production Yearbook, 1974*

Appendix
Five

MAIZE
Yield in kilograms per hectare

Country	1961–5 Average	1972	1973	1974
Algeria	960	1,000	1,000	1,000
Egypt	2,821	3,747	3,605	3,835
Iraq	1,176	1,770	1,786	1,780
Jordan	–	–	–	–
Lebanon	2,026	1,030	726	1,000
Libya	1,214	841	926	833
Morocco	796	764	487	870
Saudi Arabia	5,000	5,200	5,500	5,500
Sudan	656	970	612	667
Syria	1,043	1,303	1,328	1,462
Tunisia	–	–	–	–
Yemen Arab Republic	1,115	1,175	1,275	2,000
Yemen People's Democratic Republic	2,500	1,600	1,500	1,680

Source: Compiled and Computed from the *FAO Production Yearbook, 1974*

Appendix
Six

RICE
Yield in kilograms per hectare

Country	1961–5 Average	1972	1973	1974
Algeria	4,032	2,571	2,619	2,429
Egypt	5,307	5,208	5,430	4,898
Iraq	1,599	2,978	2,449	2,895
Jordan	–	–	–	–
Lebanon	–	–	–	–
Libya	–	–	–	–
Morocco	4,083	4,484	5,231	4,737
Saudi Arabia	–	–	–	–
Sudan	989	992	1,099	1,273
Syria	–	–	–	–
Tunisia	–	–	–	–
Yemen Arab Republic	–	–	–	–
Yemen People's Democratic Republic	–	–	–	–

Source: Compiled and Computed from the *FAO Production Yearbook, 1974*

Index